John Duncan, Charles Dixon

Birds of the British Isles

John Duncan, Charles Dixon

Birds of the British Isles

ISBN/EAN: 9783744716703

Printed in Europe, USA, Canada, Australia, Japan

Cover: Foto ©berggeist007 / pixelio.de

More available books at **www.hansebooks.com**

BIRDS OF THE BRITISH ISLES.

GREAT AUK.—P. 392.

BIRDS OF THE
BRITISH ISLES

DRAWN AND DESCRIBED BY
JOHN   DUNCAN

WITH AN INTRODUCTION BY
CHARLES   DIXON

(Reprinted from the " Newcastle Weekly Chronicle")

WALTER SCOTT, LIMITED
LONDON: PATERNOSTER SQUARE
AND NEWCASTLE-ON-TYNE
1898

THE SCOTT PRESS, FELLING, NEWCASTLE-ON-TYNE.

THIS BOOK

IS DEDICATED TO

CHARLES DIXON,

THE WELL-KNOWN NATURALIST,

AS A

MARK OF RESPECT,

BY

THE AUTHOR.

PREFACE.

THE drawings of British Birds which appear in this work were originally published in the *Newcastle Weekly Chronicle*, extending over a period of nearly ten years. Numerous correspondents of that paper expressed the wish that, when the series was completed, the drawings, with descriptions of the birds, should be published in book form. It is mainly in response to repeated suggestions of this kind that the present work is offered to the public.

The birds were drawn with the view of giving students of nature a strictly accurate representation of each specimen, with as much detail as was considered necessary for identification. Although every endeavour has been made to indicate the various markings of the feathers, bill, feet, etc., I venture to think that in no instance have I sacrificed artistic completeness for mere elaboration.

The classification is based on a simple plan, and I hope that the general reader will appreciate the system adopted. The explanatory notes that accompany each drawing are brief, but nothing material

has been omitted. In the great majority of cases
the drawing and colouration of the adult male bird
in summer plumage is given; but in many instances
the adult female and young are also described.

The Appendix contains the names of birds which
are said to have occurred in a wild state in the
British Isles, but the claims of which to be included
in the British List are doubtful.

At various times during the last ten years I
have received generous help from Mr. Richard
Howse, M.A. (Curator), Mr. Joseph Wright, many
members of the Committee, and the other officials
connected with the Natural History Museum, Newcastle-on-Tyne. Mr. John Jackson, taxidermist,
Newcastle-on-Tyne, has provided numerous and
various specimens for me to draw from, and the
members of the Ewen family, late of St. Mary's
Island, Northumberland, have also rendered valu-
able assistance in many ways. I am further deeply
indebted to Mr. Henry Hunter, Old Hartley, Northumberland; Mr. William Scott, Tynemouth; the
brothers Pow, Whitley; Mr. Amos Winship, Beal;
Mr. Edward Fenwick Wilson, Monk House, Tynemouth; Mr. Robert Wilson, Jun., Whitley; Mr.
T. W. Duncan, Monkseaton; Mr. John Fenwick
Wilson, Marden, Cullercoats; Mr. Emmanuel Fenwick Wilson, Lough House, Morpeth; Mr. John
Avery, Christon Bank; Mr. J. W. Turnbull, Corbridge; Mr. Alexander Martin, Broomhouse Lodge,

Beal; Mr. E. O. Reid, Newcastle; Mr. Thomas Lambert, Cullercoats; and numerous others, including landowners, farmers, gamekeepers, etc., for kindly assistance during a number of years past. Mr. Charles Xavier Sykes, journalist, Newcastle, has also aided me in preparing the matter for the printers. But special thanks are due to the proprietor of the *Weekly Chronicle* (Mr. Joseph Cowen), and in a minor degree to the editor of that journal (Mr. W. E. Adams), without whose cordial countenance and aid this work could not have been produced.

During the compilation of the descriptive notes, I have had occasion to make extracts from volumes written by the following admitted authorities in the ornithological world:—Charles Dixon; the late Henry Seebohm; James Backhouse, Jun., F.Z.S.; Howard Saunders, F.L.S.; R. Bowdler Sharp, LL.D.; H. Kirke Swann; Richard Lydekker, B.A.; Dr. P. M. Duncan, F.R.S.; the late Herr Gatke; the Rev. H. A. Macpherson, M.A.; and others.

Lastly, I must acknowledge my great indebtedness to Mr. Charles Dixon, author of *The Migration of Birds*, etc., etc., who, in addition to aiding me in many ways, has revised all the proof-sheets of this work.

JOHN DUNCAN.

MONKSEATON, NORTHUMBERLAND,
August 1898.

INTRODUCTION.

———•••———

ALTHOUGH the present volume requires little in the way of introducing it to an artistic and bird-loving public, I have been invited to say a few words in launching it upon what I hope will prove to be a sustained and a successful career. The books on British birds are legion; and one feels that to this fact is due the ever-increasing interest which our feathered friends and neighbours excite. It may be that many of these volumes savour too strongly of the plagiarist, the amateur, and the sciolist, to be of value (for every one must rush into print nowadays), or to advance the study of ornithology in any way, being merely tributes to the vanity of their authors, and published as such; yet, on the other hand, a welcome number contain a vast mass of novel and interesting material; they are records, too, of personal experience and observation, and therefore destined to retain a more or less permanent position in the literature of Birds.

Just a century ago, British naturalists and the British public were captivated by the first volume of Bewick's classic *History of British Birds*—a work, completed by the issue of the second volume seven years later, that gave more or less accurate pictures of all the species, and one that has retained a high place in the affections of ornithologists and artistic people generally down to the present time. To a certain extent, Mr. Duncan's volume very closely resembles the books of the immortal engraver on wood; and in some respects, as those who have the pleasure of inspecting his remarkable series of drawings will find for themselves, it is superior. The present book, so far as British birds are concerned, is practically unique. For the first time, an up-to-date manual, containing an accurate and artistic picture of every species, is placed within the reach of the multitude by reason of its very modest price. Hitherto, so far as we can recall, the cheapest with which it can be compared is published at four times the cost. Mr. Duncan's book must not be criticised from a literary point of view, its claims in that direction purposely being exceptionally small. Neither must the systematist judge it by his varying standards. It appeals to us, like Bewick's, solely through the merits of its remarkably accurate and life-like drawings of British birds—a set of pictures that we feel sure will prove of great service to any person anxious to become familiar with and able to name at a glance the four hundred or so species that are up to the present time included as British; whilst the con-

cise description of the rarer species especially should
enable the fortunate possessor of the volume to
identify any rare or strange fowl that may come in
his way.

From his childhood, Mr. Duncan has always taken
great interest in bird-life ; and this seems only natural,
for he is the son of the late Robert Duncan, the
Newcastle taxidermist, and was consequently brought
up in an ornithological atmosphere, and in a house
where the family talk was almost invariably about
birds.   At the age of fourteen John Duncan was
apprenticed to William Wailes, the well-known eccle-
siastical glass painter, with whom and with Messrs.
Wailes & Strang, he has worked for nearly forty
years.   During the early portion of that time John
Duncan studied drawing under William Bell Scott,
almost as famous as a poet as an artist.   John
Duncan became skilled in the art of painting on
glass, and many noble church windows, both in the
United Kingdom and in America, remain lasting
monuments to his talent in this direction.   During
these years John Duncan was devoting all his spare
time to the study of birds, frequently accompany-
ing his father on his ornithological excursions, and
gradually acquiring that knowledge which we feel
must have stood him in good stead during his work
upon the present volume.   Whilst in his teens, young
Duncan made many drawings of birds in oil and in
water-colours, and from that time to the present his

pencil has been seldom idle. Duncan also made
his mark as a lithographic artist and cartoonist.
For many years, we believe, he has rendered assistance to the authorities of the Hancock Natural
History Museum, Newcastle, and the grand avine
collections housed therein are indebted to him
for many rare specimens. Some ten years ago
John Duncan commenced in the *Newcastle Weekly
Chronicle* a series of pen-and-ink sketches of the
birds of the British Isles, and every week during
the whole of that lengthened period the paper has
contained some fresh example of his art. The series
now forms the present volume; and we feel sure
that not only will subscribers and others familiar
with the birds as they then appeared, be glad to
possess these drawings in a lasting and more con-
venient form, printed under the more favourable
conditions pertaining to book production, than the
hurry of the newspaper press, which can wait for
no man, but that in this form they will prove a
more worthy and lasting monument to the industry
and skill of their talented author.

CHARLES DIXON.

PAIGNTON, S. DEVON,
  October 1898.

Raven.

THE Raven (*Corvus corax*, Linnæus) is a resident species in the British Islands, although it is from various causes a diminishing species. It is also met with throughout the Palearctic Region. The adult has the entire plumage black, upper parts glossed with purple and blue; tail, nearly rounded; irides, brown; bill, legs, and feet, black. Length, about twenty-five or twenty-six inches. The young raven is duller in colouration than the mature bird.

I

Carrion Crow.

THE Carrion Crow (*Corvus corone*, Temminck) is a resident and fairly well distributed species in the British Isles.   It is said also to occur in most parts of the Western Palearctic Region, extending its range into India and also China.   The adult has the whole of the plumage black, the upper parts glossed with purple ; head, neck, and throat, tinged with green; the nostrils have a covering of bristly feathers ; legs and feet, black ; irides, brown. Length, from seventeen to nineteen inches.   The young are similar in colouration to the adult, but the gloss on the feathers is absent.

Hooded Crow.

THE Hooded Crow (*Corvus cornix*, Linnæus) breeds locally in the British Islands, but is a regular winter visitant to them. It is also found in Central and South-Eastern Europe, as well as North-East Africa and Western Asia. The adult has the head, throat, wings, and tail, iridescent black; remainder of plumage, ashy-grey; the feathers on the breast have a few dark streaks; bill, legs, and feet, black; irides, brown. Length, from seventeen to nineteen inches. The young have the head dull black, and lack the gloss of the mature bird; grey portions of plumage, duller; hind neck, dusky. This species sometimes inter-breeds with the carrion crow in Scotland.

Rook.

THE Rook (*Corvus frugilegus*, Linnæus) is a widely distributed resident in the British Isles in localities that provide
suitable trees for nesting purposes, and in districts where
a plentiful supply of food can be obtained. It is also
generally distributed in most parts of the Western Palearctic Region, ranging eastward into India. The adult
rook is easily distinguished by the purplish-black plumage
and bare grey face; bill and legs, black; irides, brown.
Length, from seventeen to twenty inches. The young are
duller in colouration, and the region round the base of the
bill is feathered.

Jackdaw.

THE Jackdaw (*Corvus monedula*, Linnæus) is not only one of the commonest resident species of the family Corvidæ, but is also one of the most widely distributed in the British Islands. It is also met with over most parts of Northern and Central Europe, but is less so in Southern Europe. It also occurs in Northern Africa. The adult has the upper plumage glossy black, with the exception of the hinder crown, neck, and sides of head, which are silvery grey; under parts, dusky black; bill, legs, and feet, black; irides, white. Length, between twelve and fourteen inches. The young are dull black in plumage, showing very little grey on the head or nape.

Common Chough.

THE Common Chough (*Pyrrhocorax graculus*, Linnæus) is a resident species with us, being found locally round the British coasts as far northwards as Skye. It is also found inhabiting certain portions of the Palearctic Region. The adult has the entire plumage glossy black, with bluish reflections; primaries and tail, slightly tinged with green and violet; bill, long and recurved, and of a scarlet colour; legs and feet, same; irides, hazel brown. Length, from fifteen to sixteen inches. The young are duller and less glossy in plumage; legs and feet, orange, but by autumn the colouration of the latter has assumed that of the adult.

Common Jay.

THE Common Jay (*Garrulus glandarius*, Linnæus), which is a resident species in the British Isles, is also met with throughout most parts of Europe, with the exception of the extreme North. As the common jay cannot very well be mistaken for any other British bird, a detailed description is not necessary. The general colouration of the plumage is vinaceous brown, with white on the rump and tail-coverts; crest, streaked with black; moustachial lines, also black; tail, brownish-black; wings, black, chestnut, and white; wing-coverts, barred with white, blue, and black; legs and feet, light brown; bill, blackish; irides, very light blue. Length, from thirteen to fourteen inches. The young are similar in colouration to the adults, but the irides are brown.

Magpie.

THE Magpie (*Pica caudata*, Gerini) is another resident species in the British Isles. It also inhabits most of the Western Palearctic Region. The adult magpie cannot well be confounded with any other of our birds of the British Isles. The adult is of a beautiful black colour on head, neck, back and breast, with reflections of violet and green ; scapulars and abdomen, white ; tail, long and iridescent with greenish-bronze, and, towards tip, purplish, shading into black ; bill, legs, and feet, black ; irides, dark brown. Length, from sixteen to eighteen inches. The young are similar in colouration, but the glossy sheen is scarcely perceptible.

Nutcracker.

JOHN DUNCAN.

THE Nutcracker (*Nucifraga caryocatactes*, Linnæus) is a rare irregular autumn migrant to the British Islands, nesting in the mountainous regions of Europe as far to the south as the Pyrenees, and occurring as far eastward as Japan. The adult male has the head dark brown; back, brown, each feather having a white triangular spot; outer rectrices, tipped with white; basal half, black; central quills with a conspicuous patch of white near the base of inner web; upper tail-coverts, brown; under parts, dark brown, with white spots on breast and belly; irides, dark brown; bill, legs, and feet, blackish. Length, about twelve inches.

Starling.

JOHN DUNCAN.

THE Starling (*Sturnus vulgaris*, Linnæus) is a resident and widely distributed species throughout the British Islands. It also occurs in summer throughout the whole of Europe, but it is only seen during winter in the countries of the Mediterranean, and ranges eastward as far as Egypt and Persia. The adult male in summer has the whole plumage black, with green, purple, and blue reflections; greater portion of feathers of upper parts, tipped with cream colour; wings and tail-feathers, blackish, margined with buffish; bill, yellow; legs and feet, brownish-red; irides, hazel. Length, from seven to eight inches. The adult female is duller in colour but is more profusely spotted. The young are ashy-brown, with pale reddish-brown margins to the wings and tail-feathers.

Rose=coloured Starling.

THE Rose-coloured Starling (*Pastor roseus*, Linnæus) is an irregular visitor to this country, occurring mostly in autumn. It is found nesting in Southern Europe, and is met with occasionally on migration in most parts of Central and Northern Europe.   In winter it is found in India, but is of rare occurrence in Africa.   The adult male in breeding attire has a crest on the head; neck, cheeks, nape, and upper part of breast, deep black, with purple reflections; wings and rectrices, black, with greenish reflections; flanks and under tail-coverts, black; back, breast, and belly, delicate rose-colour; irides, brown; bill, rose-colour, darker at base; legs and feet, brown.   Length, about eight inches. The female is duller in colour.   The young have the cheeks, chin, and upper part of throat, white; wings and tail, brown; under parts, very pale brown; mottled on flanks with darker brown.

Golden Oriole.

THE Golden Oriole (*Oriolus galbula*, Linnæus) is a rare
summer visitor to the southern portion of England. It
also occurs during the nesting season throughout Central
and Southern Europe, as well as Persia and Turkestan,
and spends the winter in Africa. The adult male has the
upper and under parts a rich golden yellow; lores and
wings, black, with a yellow alar bar; secondaries, edged
towards the tips with yellowish-white; tail, black, tipped
with yellow more broadly on the outermost feathers; bill,
dull red; feet, leaden grey; irides, red. Length, about
nine inches. The adult female and young resemble the
male in colouration, but are greyish-white below, with
streaks of blackish on the throat and breast.

Common Crossbill.

THE Common Crossbill (*Loxia curvirostra*, Linnæus) is an uncertain winter visitor to the British Isles, although some remain with us. It is also found in summer in most parts of Europe and Northern Asia. The general colour above of the adult male is dull crimson, becoming brighter on the rump and chest; wings and tail, brown, with pale margins to wing-coverts; vent, white; irides, hazel; bill, legs, and feet, dark brown. Length, about six inches. The adult female is mostly greyish-brown, tinged on the upper parts and breast with greenish-orange; rump, brighter. The young are dull in colouration similar to the adult female, but are streaked on the upper and under parts.

American White=winged Crossbill.

THE American White-winged Crossbill (*Loxia leucoptera*, Gmelin) is a rare nomadic autumn migrant to the British Isles, and is also found in Northern North America, from Alaska to Labrador and Newfoundland. This crossbill is said to have a more slender bill than the European white-winged crossbill.

European White=winged Crossbill.

JOHN DUNCAN

THE European White-winged Crossbill (*Loxia bifasciata*, Brehm), which inhabits Northern Russia as well as Siberia, is a rare wandering migrant to the British Isles during autumn and winter. The adult male of this crossbill is said to be distinguished from the common crossbill by the white bands on the wings. Length, about six inches.

parrot Crossbill.

JOHN DUNCAN

THE Parrot Crossbill (*Loxia pityopsitacus*) has occurred at various times in the British Isles, and is found inhabiting Scandinavia and Northern Russia. This crossbill is now considered to be merely a race of the common crossbill.

Pine Grosbeak.

THE Pine Grosbeak (*Loxia enucleator*, Linnæus) is an extremely rare winter migrant to the British Isles. It is a resident in the Palearctic and Nearctic Regions. The adult male has the general colouration of the plumage dull slate-grey; wings, dark brown, with rosy margins to the feathers, getting broader on inner secondaries; lores, black; remainder of head, crimson; rump and upper tail-coverts, mostly red; rectrices, black tinged with grey, and margined with paler; under parts, crimson, shading into ashy-white on lower belly and under tail-coverts; irides, bill, legs, and feet, dark brown. Length, about eight inches. The female shows no crimson colour.

2

Scarlet Rose Finch.

THE Scarlet Rose Finch (*Carpodacus erythrinus*, Linnæus) is an extremely rare visitor to the British Islands. It inhabits the Palearctic Region, and is found westward in Europe to Finland and Poland. In Western Europe it is a mere straggler. The adult male has the back brown, washed with rose; wings and rectrices, pale brown, margined faintly with crimson; rump, crimson; under parts below breast, white, shading into creamy white on under tail-coverts; crown of head, chin, throat, and breast, crimson, the crown being the darkest; irides, legs, and feet, brown; bill, brown tinged with yellow. Length, about five inches. The adult female is, above, mostly olive-brown; wings and rectrices, dark brown; innermost secondaries margined with whitish, and coverts tipped with whitish; fore-neck and breast, buffish, with streaks of dusky; belly, white.

Hawfinch.

THE Hawfinch (*Coccothraustes vulgaris*, Pallas) is a fairly
common resident in certain localities in the British Islands.
It is also met with throughout the greater part of the
Palearctic Region. The adult male hawfinch in summer
plumage can readily be identified by its bulky head, strong
bill, and the secondary feathers being notched. The general
plumage is brown, above; wings and tail, black and white;
under parts, yellowish-brown; bill, leaden blue, dull black
at tip; legs and feet, pale fleshy-brown; irides, greyish-
white. Length, about seven inches. The adult female is
similar in colouration, but duller.

Bullfinch.

THE Bullfinch (*Pyrrhula vulgaris*, Temminck) is a resident species in the British Isles, being also met with in Europe west of Central Russia and south of Scandinavia. Examples occasionally occur in the southernmost countries of Europe. The adult male has the crown of the head, base of bill, and throat, velvety blue-black; nape and mantle, smoke-grey; larger wing-coverts have whitish tips, thereby producing a transverse bar; wings and tail, bluish-black; rump, white; under tail-coverts, white; rest of under parts, dull vermilion; bill, black; legs and irides, dark brown. Length, about six inches. The female on the upper parts is brownish-grey; underneath, warmish brown. The young have no black on the head, and the wing bar is buffish-white.

Greater Bullfinch.

THE Greater Bullfinch (*Pyrrhula major*, C. L. Brehm) is a rare visitor to the British Isles, nesting in Scandinavia and North-Eastern Europe generally, and occurring during winter in South-Eastern Europe. The adult male closely resembles *Pyrrhula vulgaris*, but in some examples the colours are more brilliant; it is also larger in size.

House Sparrow.

JOHN DUNCAN

THE House Sparrow (*Passer domesticus*, Linnæus) is a well-known resident and generally distributed species throughout the British Isles, and also over the whole of the Palearctic Region. The adult male in summer has the top of the head and rump slaty-grey; chin, throat, breast, and lores, black; above the eyes, a thin white streak; mantle, ruddy brown, with dark centres to the feathers; wings and tail, dark brown, margined with paler brown; white tips to lesser wing-coverts; cheeks, ashy; under parts, greyish-white; bill, black; legs and feet, brownish; irides, brown. Length, about six inches. In the adult female the black on the throat and breast are absent.

Tree Sparrow.

THE Tree Sparrow (*Passer montanus*, Linnæus) is a generally distributed, but local, species in the British Isles. It is also found throughout most parts of the Palearctic Region, and in the Indian Region, reaching as far southwards as Java. The adult male in breeding plumage has the crown and nape chocolate-brown; throat and fore-neck, black; lores, and a streak behind eyes, black; ear-coverts, ashy-white, with a black patch on the lower parts; sides of neck, white; upper parts and tail closely resemble the house sparrow, with the exception that the median and greater wing-coverts have white tips, which form two bars; under parts, ashy; bill, black; legs, pale brown; irides, brown.   Length, about five inches.   The adult female is similar to the male in colouration.

Chaffinch.

THE Chaffinch (*Fringilla cœlebs*, Linnæus), which is a resident species in the British Isles, is also found in most parts of Europe, but is rarely met with in North-Western Africa. The adult male in summer has the forehead black; crown and nape, slaty grey; back, reddish-brown; rump, sap-green; scapulars, grey, tipped with brownish; quills, dusky, with narrow margins of white; lesser and median wing-coverts, white; greater coverts, blackish, with broad white tips; central tail-feathers, grey; remainder, dark brown; two outermost rectrices on either side, white on the inner margins; sides of head, and lower part of plumage, reddish-brown, shading into white on abdomen; bill, dusky; irides, hazel; legs, dull brown. Length, about six inches. The female's head and back are olive-brown; cheeks, throat, and breast, pale yellowish-grey.

Brambling.

THE Brambling (*Fringilla montifringilla*, Linnæus) is a common autumn and winter migrant to the British Isles, and is also found throughout most of the Palearctic Region, nesting in the Far North. The adult male in nuptial plumage has the general colour above, also sides of neck, ear-coverts, and cheeks, glossy black; lesser wing-coverts, orange-chestnut; median coverts, white; greater coverts, with white tips; rump, white; a white speculum on the wing; tail, black; outermost feathers, partly white; chin and throat, ochreous; remainder of under parts, ochreous, tinged with red, gradually turning into whitish on belly and under tail-coverts; sides, spotted with black; irides, dark brown; bill, black; legs and feet, brown. Length, from five to six inches.

Greenfinch.

THE Greenfinch (*Fringilla chloris*, Linnæus) is a resident
species in the British Islands.   It also occurs in most parts
of Europe (with the exception of the Spanish Peninsula)
north to southern Scandinavia, as well as in Turkestan.
The adult male in summer has the forehead greenish-
yellow ; lores, dusky black ; a bright yellow stripe over each
eye ; cheeks, and under parts of body, bright yellow, washed
slightly with ashy ; crown, neck, and back, olive green ;
feathers margined with greyish ; rump, yellowish ; wings,
deep brown, with dull grey tips ; primaries, edged with
bright yellow at their bases ; tail, dark brown, central
feathers margined with grey ; remainder with entirely yellow
bases ; flanks, greyish ; bill, light flesh colour ; irides, hazel.
Length, about five inches.   The adult female is much
duller in colouration than the male.

Goldfinch.

THE Goldfinch (*Fringilla carduelis*, Linnæus) is a generally distributed species in the British Islands.   It is also
a resident throughout most parts of Europe, with the
exception of the extreme north.   In Northern Africa and
eastward into Persia it is also a resident.   The adult male
has the forehead and upper throat crimson ; crown of the
head and a band behind the ear-coverts, black ; cheeks
and lower throat, white ; upper parts, pale chocolate-brown ;
greater wing-coverts, brilliant golden yellow ; primaries,
black with white tips ; base of primaries, banded with bright
yellow ; tail, black, tipped with white ; outer tail-feathers
subterminally having white blotches ; underneath, white ;
sides of breast and flanks, tinged with wood-brown ; bill,
pale brown, tip darker ; irides, legs, and feet, brown.
Length, about five inches.

Siskin.

THE Siskin (*Fringilla spinus*, Linnæus) is generally considered a winter visitor to the British Isles, although instances are recorded of its having nested in most of our English counties. It also frequents the temperate parts of the Palearctic Region. The adult male has the crown and throat black; back, dullish-green, with streaks of dark brown; rump, yellow; cheeks and ear-coverts, olive-green; wings, blackish, with yellow bars and pale yellowish margins; tail-feathers, blackish, yellow at base with the exception of centre ones; under parts, yellow, shading into white on abdomen; flanks, yellowish, with dusky streaks of black; irides, brown; bill, dusky; feet, light brown. Length, about four inches. The adult female has the crown streaked with dusky brown; wings, rump, and tail show very little yellow; under parts, yellowish-white, with dusky streaks.

Serin Finch.

JOHN DUNCAN.

THE Serin Finch (*Serinus hortulanus*, Koch) is a rare abnormal spring and autumn migrant to the British Islands. It is a resident in most parts of Central and Southern Europe. In Asia Minor it is plentiful, and it is also met with in Palestine. The adult male in summer has the general colour above, greenish-olive, with streaks of dark brown; forehead and cheeks, yellow, tinged on the latter with green; wings and rectrices, dusky, margined with pale yellowish; throat, breast, and rump, bright yellow; middle of abdomen, white; flanks, streaked with dark brown; irides, dark brown; bill, dark brown; legs and feet, brown. Length, about four inches. The adult female is duller in colouration, and shows less yellow. The young resemble the adult female, but the yellow is absent.

Canary.

JOHN DUNCAN.

THE Canary (*Serinus hortulanus canarius*, Linnæus) is an accidental spring and autumn visitor to the British Islands, and a resident in the Canary Islands, Madeira, and the Azores. The adult male wild canary has, it is said, "the crown, yellowish-green, narrowly streaked with blackish-brown; back, and upper tail-coverts, blackish-brown, margined broadly with grey, and marked with olive-green; forehead and rump, bright yellowish-green; wing-coverts, blackish-brown, tipped and margined with olive-green. The whole of the under parts are golden yellow, shading into almost white on the under tail-coverts; the flanks are striped with blackish-brown; bill, light brown; legs, feet, and claws, fleshy brown; irides, hazel." The adult female is described as being duller in colour, and shows less yellow.

Linnet.

JOHN DUNCAN.

THE Linnet (*Linota cannabina*, Linnæus) is a generally distributed species in the British Isles, but becomes scarcer in some parts of Scotland, and is said not to occur in the Shetlands. It is also a resident over most parts of the Western Palearctic Region, with the exception of the extreme north. The adult male in summer plumage is, above, reddish-brown, streaked with black; forehead and centre of crown, crimson; breast, crimson; ear-coverts, ashy-grey; primaries and tail, blackish, margined with white; throat, dusky; breast, crimson; abdomen, dull white; flanks, fawn-brown; bill, legs, and feet, brown; irides, brown. Length, about five inches. The adult female closely resembles the male, but the red on crown and breast is wanting.

Twite.

THE Twite (*Linota flavirostris*, Linnæus) is a resident in the British Islands, but is stated to be only a winter visitant to the south. It nests in Norway, and is also met with on migration in most parts of Europe. The coloura- tion of the plumage of the adult male in summer is, above, dark brown, marked with tawny-brown; sides of head and also body, marked with brown; rump, rose red; quills, dusky, a few primaries having narrow white margins, and secondaries tawny margins; tail-feathers, dusky, with white margins to the outer ones; chin and throat, unspotted tawny-buff; sides of head and body, marked with brown; abdomen, white; bill, yellowish, tip darker; irides, brown; legs and feet, brown. Length, about five inches. The female is distinguishable by having no red on rump.

Lesser Redpole.

THE Lesser Redpole (*Linota rufescens*, Vieillot) is a common species in the British Isles during winter, but is locally distributed in summer.   It is also found in the more northern portions of both hemispheres, nesting mostly in regions where birch trees abound, and in winter straying southward.   The adult male in spring plumage has the throat and lores black; crown and forehead, deep crimson; upper parts, dark warm brown; rump, pink; wing bands, rufous buff; breast, carmine red; remainder of plumage closely resembles that of the mealy redpole (*Linota linaria*); irides, brown; bill, brownish horn colour; base beneath, yellowish; legs, dark brown.   Length, about four inches.   The female, as a rule, has no red on the rump or breast,

Mealy Redpole.

THE Mealy Redpole (*Linota linaria*, Linnæus) is an autumn migrant to the British Isles. It inhabits the more northern portions of both hemispheres, and nests principally in the birch region, and in winter moves southward. The adult male in summer has the nape, back, and scapulars dark brown, with whitish margins to some of the feathers; primaries, dusky; innermost secondaries, greater and middle wing-coverts, with whitish tips; rump, white, tinged with pink and streaked with brown; rectrices, dusky, with narrow margins of whitish; crown of head, crimson; chin, black; breast, suffused with carmine; middle of abdomen, whitish; sides, streaked with dark brown; irides, brown; bill, yellowish-brown; legs and feet, brown. Length, five inches. The adult female in summer shows no red on the breast or rump. In winter, upper parts of the adult are lighter and the feathers show no red on the breast.

Greenland Redpole.

THE Greenland Redpole (*Linota linaria Hornemanni*, Holboell) is an extremely rare abnormal winter visitor to this country, and so far only one example has been procured. It is met with in Greenland and Eastern North America. This redpole can readily be distinguished from its congeners by the excessive whiteness of the plumage.

Coues' Redpole.

COUES' Redpole (*Linota exilipes*, Coues) is an exceedingly rare abnormal visitor to the British Isles.

Holboell's Redpole.

HOLBOELL'S Redpole (*Linota Holboelli*, Brehm) is a rare abnormal migrant to the British Isles, and inhabits North Europe from Scandinavia to Eastern Siberia. This redpole is described as being similar to the mealy redpole, "but larger in size, and having a very much larger bill."

Snow Bunting.

JOHN DUNCAN.

THE Snow Bunting (*Plectrophenax nivalis*, Linnæus) is mainly a winter visitor to the British Isles, but instances are recorded of its having nested on some of the Scottish mountains. It is a summer visitant to the Arctic Regions, and migrates southwards in the autumn through most of the northern countries in Europe, Asia, and North America. The adult male in summer has the head, neck, upper wing-coverts, secondaries, and under parts, white; back and scapulars, black, the back on lower part usually mottled with white; primaries and tertiaries, black; six central tail-feathers, black; remainder, white; irides, brown; bill, legs, and feet, black. Length, about six inches.

Lapland Bunting.

THE Lapland Bunting (*Calcarius lapponicus*, Linnæus) is a rare irregular autumn migrant to the British Isles. It nests throughout the Circumpolar Regions north of the Arctic Circle, and in autumn migrates through Central Europe and Asia. The adult male in summer has the head, throat, and upper breast black, with the exception of a whitish streak, which commences behind the eye and passes at the back of the auriculars, thus forming a patch of white on the sides of the neck; a collar of bright chestnut reaches from the occiput on to the upper back; remainder of upper parts, wing-coverts, innermost secondaries, and middle rectrices, brownish-black margined with buff and white; primaries, with pale margins; under surface of body, pale buffish, striped on flanks with black; irides, brown; bill, yellow; tip, black; legs and feet, black. Length, about six inches. The adult female lacks the black head and rufous collar.

Reed Bunting.

THE Reed Bunting (*Emberiza schœniclus*, Linnæus) is a resident species in the British Islands, and also in suitable localities throughout the Palearctic Region. The adult male is rufous colour above, with black streaks and pale margins to the feathers; head and throat, deep black; a white moustachial line runs from the base of the bill and joins the collar, which, like the breast, is white; lower back and rump, pale grey, with black streaks; scapulars and lesser wing-coverts, warm reddish-brown, with dark centres on the former; tail, dark brown, two outer pairs of rectrices having oblique white patches; flanks, streaked with dusky brown; irides, brown; bill, brown above, paler below; legs and feet, brown. Length, about six inches. The adult female lacks the black head and throat.

Corn Bunting.

THE Corn Bunting (*Emberiza miliaria*, Linnæus) is a
resident species in the British Isles, and is also met with
throughout most parts of Europe, with the exception of
the Far North. It is also found ranging eastward through
Central Asia. The colouration of the adult male in
summer is, above, hair-brown, with dark centres to the
feathers; lower back and rump not so distinctly marked;
wing-coverts, dark brown, margined with buff; primaries,
dullish brown; tail, dark brown, with pale margins;
moustachial line composed of dark brown spots; throat
and under parts, buffish-white, profusely spotted on the
breast and streaked with brown on the flanks; bill,
yellowish-brown; irides and legs, brown. Length, about
seven inches. Both sexes are alike.

Cirl Bunting.

THE Cirl Bunting (*Emberiza cirlus*, Linnæus) is a very local species, being principally found in the South of England. It is also met with in Western Europe, Southern Europe, Asia Minor, Algeria, and North-Western Africa. The colouration of the adult male is, above, chestnut, with black streaks. Head and hinder part of neck, olive, streaked with black; rump, greenish-olive, with streaks of dusky; eye-stripe and one below, yellow; lores and ear-coverts, black; throat, black, succeeded by a collar of pale sulphur colour; underneath is a band of olive-grey, followed by stripes of reddish-brown running down the flanks; abdomen, pale dull yellow; irides, brown; bill, leaden colour; legs and feet, pale brown. Length, about six inches. The female has no black or yellow markings on the face.

Yellow Bunting.

THE Yellow Bunting (*Emberiza citrinella*, Linnæus) is a resident in the British Islands. It is also a resident throughout Europe and in Asia, reaching as far eastward as Turkestan. The adult male in summer has the crown yellow, tinged with green on the sides; eye-stripe, yellow; above, brown, feathers having black centres; rump and tail-coverts, chestnut; tail-feathers, mostly dark brown, with elongated white patches on the two outermost pairs; wings, dark brown; primaries, dusky brown, margined narrowly with yellowish; beneath, yellow, tinged on chest with dull olive; flanks, streaked with blackish-brown; irides, dark-brown; bill, horn colour; legs and feet, light brown. Length, between six and seven inches. The female is duller in colouration than the male.

Ortolan Bunting.

John Duncan

THE Ortolan Bunting (*Emberiza hortulana*, Linnæus) is a rare irregular spring and autumn migrant to the British Isles; it is also a visitor in summer to temperate Europe and Central Asia, and in winter moves into Northern Africa and India.   The adult male has the general colour of the upper parts reddish-brown; feathers on mantle and back, streaked with black; head, grey, with a tinge of greenish-yellow; a ring round eye, lores, chin, and throat, yellow; sides of neck and band across chest, yellowish-green; under parts, cinnamon; irides, brown; bill, red; legs and feet, reddish.   Length, about six inches.   The adult females resemble the male, but are lighter in colour, and the chest has mottlings of dusky brown.

Black=headed Bunting.

JOHN DUNCAN

THE Black-headed Bunting (*Emberiza melanocephala*, Sco-poli) is a very rare irregular spring and autumn visitor to the British Isles. In summer it is exceedingly common in South-Eastern Europe; it spends the winter in India. The adult male in nuptial dress has the wings brown; coverts, tipped with pale brown; innermost secondaries, broadly edged with pale brown; rectrices, brown; head, lores, and auriculars, black; rest of upper parts, chestnut, extending to the sides of breast; under parts, rich yellow, tinged with rufous on sides of body; irides, dark brown; bill, grey; legs, pale brown. Length, about six inches. The adult female lacks the black on the head, and plumage generally is brownish and yellowish. In winter the whole plumage has a dingy appearance. The young in first plumage are very similar to the adult female.

Little Bunting.

JOHN DUNCAN.

THE Little Bunting (*Emberiza pusilla*, Pallas) is an extremely rare autumn migrant to the British Isles. In summer it is found in North-Eastern Europe and Siberia, and spends the winter in India and China. The adult male in nuptial plumage has the upper parts rufous-brown, conspicuously streaked with black; tail, dark brown, edged with paler brown, except two outermost feathers which have elongated patches of white on inner webs; wings, barred with buff; head, bright chestnut, black band on each side of the crown; chin and upper throat, rufous-chestnut; under parts, white, with black streaks on breast and flanks; irides, bill, legs, and feet, dark brown. Length, five inches. The adult female in summer is similar to the male, but the plumage is duller.

Rustic Bunting.

JOHN DUNCAN.

THE Rustic Bunting (*Emberiza rustica*, Pallas) is a rare irregular autumn migrant to the British Isles.   It also inhabits Northern Russia and Siberia, in winter migrating westward and southward.   The adult male in summer has the back and scapulars reddish-brown, with mottlings of black, and bordered broadly with buff; rump, rich chestnut; crown and sides of head, black; eye-stripe and throat, white; band across chest, rich chestnut; rectrices, deep brown; outermost pair of feathers, mostly white; breast and belly, white; flanks, streaked broadly with rich chestnut; irides, dark brown; bill, dark brown above, yellowish beneath; legs and feet, yellowish flesh colour.   Length, about five inches.   The colouration of the female is generally duller than that of the male.

Brandt's Siberian Bunting.

BRANDT'S Siberian Bunting (*Emberiza cioides*, Moore) is a partially migratory species, an example being taken near Flamborough. It is found in Northern China.

Sky-Lark.

JOHN DUNCAN

THE Sky-Lark (*Alauda arvensis*, Linnæus) is a generally distributed resident species throughout the British Islands. It is also a resident throughout nearly the whole of the Palearctic Region. The adult male in summer plumage is of various shades of brown, here and there tinged with yellowish-buff. The feathers on the top of the head form a dark brown crest, with paler edges. Upper parts, brown, with darker centres; eye-stripe, buffish; tail-feathers have dark brown centres and edged with tawny; outer pair, chiefly white; second pair, white on outer webs; below, tinged with tawny-buff or rufous, boldly spotted on sides of breast and throat with dark brown ; irides, dark brown ; bill, brown ; legs and feet, pale yellowish-brown.   Length, about seven inches.

Wood=Lark.

THE Wood-Lark (*Alauda arborea*, Linnæus) is a somewhat widely distributed resident species, but is very local.   It is said to be unequally distributed over the Western Palearctic Region, south of latitude 60° N.   The adult in summer resembles the sky-lark in the general colouration of the plumage and crest, but can readily be distinguished from the latter by its smaller size, shorter tail, and the stripe over the eye, which is very broad, and buffish-white in colouration, and reaches to the nape of the neck.   Irides, dark brown; bill, brown; legs and feet, pale brown.   Length, about five inches.   The adult female is similar in colouration to the adult male.   The young are more rufous in colour than the adults, and the feathers have light buff tips.

Short=toed Lark.

JOHN DUNCAN

THE Short-toed Lark (*Calandrella brachydactyla*, Leisler) is a rare abnormal autumn migrant to the British Islands. It is a resident in Southern Europe and Northern Africa, and has been met with as far eastward as India. The adult male in nesting plumage has the upper parts chiefly light sandy-brown, with blackish centres to the feathers, those on the head narrow, and on rump, obscure; under parts, pale ochreous-white; breast, washed with buffish-brown; with a few large streaks on the sides of the breast; tail-feathers, dark brown; centre webs of outermost pair, mostly creamy-white; next pair with terminal part of outer web, creamy-white; central rectrices, margined with pale sandy-grey; irides, brown; bill, brown; legs and feet, brownish. Length, about five inches. The adult female is similar to the male.

Crested Lark.

JOHN DUNCAN

THE Crested Lark (*Galerita cristata,* Linnæus) is a rare irregular autumn migrant to the British Islands, but is a resident species in most parts of Continental Europe and Asia. The adult male has the upper parts of the plumage brown, with centre of feathers blackish; nape and upper back, washed in parts with pale ochreous brown; rump and upper tail-coverts, buff; tail-feathers, brown, with paler margins; superciliary stripe, buff; crest, which is conspicuous, has the feathers buff in colour with dark brown centres; lower parts, buffish-white; sides of throat and chest, spotted with brown; flanks, streaked with dark brown; irides, hazel; bill, brown, paler beneath; legs and feet, light brown. Length, about six inches. The adult female is similar to the male.

White=winged Lark.

JOHN DUNCAN

THE White-winged Lark (*Melanocorypha sibirica*, Gmelin) is a very rare autumn visitor to the British Isles, but in parts of Russia and Central Asia it is a resident, and is met with occasionally in Austria and Italy. The adult male in summer has the crown, lesser wing-coverts, and upper tail-coverts, chestnut; tail, brown, margined with white on the inner webs; upper parts, brown, margined with russet, and shading into lighter on nape; secondaries, white, dark at bases; primaries, dark brown, with white tips; under parts, dull white, tinged with buffish on breast and shading into brownish on flanks, the former finely spotted and the latter streaked with brown and rufous; irides, brown; legs and feet, brown; bill, deep horn colour, pale yellowish beneath. Length, about seven inches. The adult female is similar to the male, but duller in colouration.

4

Calandra Lark.

THE Calandra Lark (*Melanocorypha calandra*, Linnæus) is an exceedingly rare autumn migrant to the British Isles, but is a permanent resident in the South of France and in Spain, also in parts of Italy and Greece, and is very numerous in Turkey and North Africa. The adult is greyish-brown on the upper parts, with dark centres to the feathers ; under parts, white, washed with tawny, and finely streaked with dark brown. On each side of the neck is a conspicuous patch of black. Length, about six inches.

Shore=Lark.

THE Shore-Lark (*Otocorys alpestris*, Linnæus) is a regular autumn migrant to the British Islands, and is also found in Northern Europe, Asia, and America, and during migration occurs in most parts of Europe. The adult male has the crown, ear-tufts, lores, and lower neck band, black; forehead, superciliary stripe, chin, and upper throat, yellowish; general colour of upper parts, greyish-brown, tinged with pinkish-brown on nape, lesser wing-coverts, and upper tail-coverts; feathers on back have dark centres; centre tail-feathers, ashy, centred with black; remainder, black, outer one edged with white; under parts of body, white, washed on flanks with pale brown; irides, dark brown; bill, slate colour; legs and feet, blackish. Length, about seven inches. The adult female has no ear-tufts.

Pied Wagtail.

THE Pied Wagtail (*Motacilla alba Yarrellii*, Gould) is a resident in, and is widely and generally distributed over, the British Isles during the nesting season, occurring also in Western Europe; elsewhere it is only a summer visitant; it spends the winter in the western portions of North Africa. The adult male in summer has the top of the head, nape, part of the shoulders, chin, neck, throat, and upper breast, rich black; forehead, sides of the face, and patch on sides of neck, white; mantle, rump, and wing-coverts, black; the latter having white margins to the feathers, forming a double bar; primaries, blackish; inner secondaries, with white margins on outer edge; abdomen, white; sides and flanks, tinged with blackish; tail, black; outermost feathers, white on outer webs; irides, dark brown; bill, legs, and feet, black. Length, seven inches,

White Wagtail.

THE White Wagtail (*Motacilla alba*, Linnæus) is a regular spring visitant to the British Isles, and has been on several occasions identified as having nested in this country. It is a resident species in some parts of Southern Europe, and spends the winter in North Africa. The adult male in summer has the chin, throat, and breast, black; auriculars, and patch on sides of neck, white; above, grey; wing-coverts, with black bases and edged with white; crown and nape, black; rump and upper wing-coverts, ashy-grey, tipped with white; median ones also with white tips; primaries, blackish; under parts, white; sides, ashy-grey; tail-feathers, black in centre; outer ones, white, edged with black inwardly; irides, dark brown; bill, legs, and feet, black. Length, about seven inches. The female is similar to the male, but the black on the crown is mixed with ashy.

Grey Wagtail.

THE Grey Wagtail (*Motacilla sulphurea*, Bechstein) is a resident species pretty generally distributed throughout the British Islands, but is more local in the South of England. It is also a resident in Central and Southern Europe, ranging across Central Asia; and is to be met with in winter in India, Persia, and Northern Africa. The adult male in summer is bluish-grey above; rump, yellow; eye-stripe, white; moustachial line, white; throat, black; under parts, bright yellow; under tail-coverts, richer yellow; wings, greyish-black; coverts, tipped with whitish; tail, blackish-brown; outer rectrices, white; irides, brown; bill, legs, and feet, brown. Length, between seven and eight inches. The adult female is similar, but colours not so brilliant. The young are browner in colour than the female, and the eye-stripe is buffish.

Blue=headed Wagtail.

THE Blue-headed Wagtail (*Motacilla flava*, Linnæus) is an irregular spring and autumn visitor to the British Isles. It is mostly met with on the South Coast, and has nested in the North of England. It also nests in Central Europe and Central Asia, and migrates to Africa in winter. The adult male in summer has the crown and nape bluish-grey ; distinct white eye-stripe ; back, deep yellowish-olive ; wing-coverts, dark brown, with yellowish-white tips, which form a double bar ; secondaries, margined with yellowish-white ; primaries, dark brown ; tail, dusky-brown ; two outer tail-feathers on each side, mostly white ; chin and underneath lores, white ; under parts, bright yellow ; irides, brown ; bill, legs, and feet, dark brown. Length, about six inches. The female is duller and fainter in colour than the male.

Yellow Wagtail.

THE Yellow Wagtail (*Motacilla raii*, Bonaparte) is a widely distributed common summer visitor to the British Isles. It also nests in Central and Western Europe, and spends the winter in Africa. The adult male in summer has the general colour above, olive-yellow; under parts, bright yellow; superciliary stripe, yellow; wings, dark brown; wing-coverts and secondaries, tipped and margined with yellowish-white; tail-feathers, blackish brown, with the exception of the two outer pairs, which are white; irides, dark brown; bill, legs, and feet, black. Length, about six inches. The adult female is similar to the adult male, but the colours are duller, and show a brown tinge on the back. The young somewhat resemble the female, but have the throat and chest light fawn colour; there are also brown spots on the chest; remainder of under parts, yellow.

Tree Pipit.

THE Tree Pipit (*Anthus trivialis*, Linnæus) is a summer visitor to the British Islands, but it is comparatively rare in Scotland, and is exceedingly scarce in Ireland. It visits Central and Northern Europe and Asia during summer, spending the winter in Africa, Persia, and India. The adult in summer has the upper parts dark brown, margined broadly with buffish-brown; wings and most of tail-feathers, dark brown, edged with paler brown; about one-half of the outer pair of rectrices, white; chin, whitish; throat, buff; beneath, pale buff; abdomen, white; sides of throat, chest, and flanks, spotted and streaked with dark brown; irides, dark brown; legs and feet, pale brown. Length, about six inches. The female is similar to the male, but the streaks on the under parts are narrower. This species is readily distinguished from *Anthus pratensis* by its shorter and much curved hind claw.

Meadow Pipit.

THE Meadow Pipit (*Anthus pratensis*, Linnæus) is a common resident species throughout the British Isles, but is subject to partial migration in winter. It also breeds in most parts of Europe, and winters in Africa. The adult in summer is olive-brown above, the feathers having black centres, those on the mantle margined with whitish; wing-coverts and secondaries, brown, edged with buffish-white; primaries, dark brown, with exterior margins of yellowish-green; eye stripe, yellowish-buff; throat and breast, tawny-buff, the latter closely streaked with blackish-brown; flanks, tinged with olive and streaked broadly with black; lower parts, buffish; tail, dark brown; outer tail-feathers, mostly white; irides, dark brown; bill, blackish; legs and feet, light brown. Length, about six inches.

Alpine Pipit.

JOHN DUNCAN

THE Alpine Pipit (*Anthus spipoletta*, Linnæus) is a rare irregular autumn migrant to the British Isles. It nests in the mountain regions of Central Europe, also in parts of Spain; and on migration occurs in other portions of Europe, passing the winter in Northern Africa. The adult male in summer is olive-brown above; paler on rump; feathers of upper parts, darker in centre; eye-stripe, creamy-white; wings and tail, dark brown, margined with paler; the outermost tail-feathers, mostly white; chin, centre of abdomen, and under tail-coverts, pale ochreous-white; remainder of under parts, pale buff tinged with chestnut; flanks, olive-brown; irides, dark brown; bill, blackish, paler beneath; legs and feet, blackish. Length, about six inches.

Rock Pipit.

THE Rock Pipit (*Anthus obscurus*, Latham) is a resident
in the British Islands; it also occurs as a summer visitant in
Northern Europe, and in winter is found in Central and
Southern Europe. The adult male in summer plumage is
olive-brown above, the feathers having dark centres; wings
and tail, dark brown, with paler edges to the former and
olivaceous edges to the latter; eye-stripe, buffish-yellow;
throat, dull white; beneath, buffish-yellow, feathers closely
mottled with dark brown centres; irides, brown; bill, legs,
and feet, brown; the hind claw is much curved. Length,
about six inches. The adult female resembles the male,
but underneath is less spotted. The young are darker in
colouration, and the under parts are more heavily marked.

Tawny Pipit.

JOHN DUNCAN

THE Tawny Pipit (*Anthus campestris*, Linnæus) is a rare and irregular autumn migrant to the British Islands. It breeds in Central as well as Southern Europe, passing the winter in Africa and India. The adult male has the upper parts sandy-brown, with darker centres; rump, sandy-brown; central tail-feathers, dark brown, margined with lighter; wings, dark brown; under parts, buffish-white; darkest on breast, the latter faintly spotted; irides, dark brown; bill, dark brown; lower mandible, light brown; legs and feet, pale brown. Length, about six inches. The female closely resembles the male. This species is subject to much variation in colour of plumage.

Richard's Pipit.

RICHARD'S Pipit (*Anthus Richardi*, Vieillot) is an irregular autumn migrant to the British Islands, and is also found inhabiting Central Asia, and during migration is met with accidentally in various parts of Europe, wintering in Southern Asia and North-Eastern Africa. The adult male in nuptial plumage has the upper parts mostly of a yellowish-brown tint, with darker centres to the feathers of the head and back; eye-stripe, whitish; two outermost tail-feathers, nearly white; under parts, dull white, tinged on breast and sides with reddish-brown, and streaked with dark brown; irides, hazel; bill, dark brown, paler beneath; legs and feet, pale brown. Length, from seven to eight inches. This pipit can easily be distinguished from other pipits by its large size and developed hind claw.

Red=throated Pipit.

THE Red-throated Pipit (*Anthus cervinus*, Pallas) is an extremely rare spring migrant to the British Islands, visiting Arctic Europe and Asia during summer, and wintering in North-Eastern Africa and the South of Asia.   The adult male in summer has the upper parts sandy-brown, tinged with olive; centre of feathers with bold markings of dark brown; wings, dark brown, margined with paler; above and below the eye, sides of neck, throat, and chest, rusty red; tail, dark brown, with paler margins; two outermost feathers principally white; under parts, pale buffish-brown, thinly streaked with dark brown; irides, brown; bill, brown; base beneath, yellowish; legs and feet, light brown.   Length, about six inches.   The adult female in summer is similar to the male, but lacks the rusty red on the chest.

Common Creeper.

THE Common Creeper (*Certhia familiaris*, Linnæus) is a more or less commonly distributed resident species in the British Islands, and throughout most parts of Europe. The adult has the crown of the head dark brown, with whitish centres to the feathers; stripe over eye, whitish; nape and mantle, rufous-brown, streaked in centre of feathers with buffish-white; wings, brown, barred with paler brown; secondaries, tipped with whitish; under parts, silvery-white, tinged with buff on the flanks; tail, graduated with stiff points, and dull reddish-brown in colour, the feathers having yellowish shafts; irides, hazel; bill, strongly decurved, dark brown; beneath, yellowish; legs and feet, pale brown. Length, about five inches.

Wall=Creeper.

THE Wall-Creeper (*Tichodroma muraria*, Linnæus), which is an exceedingly rare and irregular spring and autumn migrant to the British Isles, is a resident in the mountainous parts of Southern Europe, and is also met with in some portions of the Pyrenees and Sierra Nevada.   The adult male in summer has the upper parts pearl-grey; top of head, tinged with brown; rump and upper tail-coverts, blackish; tail, black, tipped with grey and white; wing-coverts and basal halves of outer webs of primaries, rich crimson; rest of primaries, dark brown, with conspicuous

5

spots of white on longest quills, the latter tipped with white; belly and under tail-coverts, black; irides, brown; bill, legs, and feet, black. Length, about six inches. The adult female in summer is similar to the male.

Firecrest.

THE Firecrest (*Regulus ignicapillus*, Brehm) is an abnormal autumn and winter migrant to the British Isles, and is a resident in most of temperate Europe, but does not travel so far northwards as the golden-crested wren. The adult male is "above, olive-green; brighter on sides of neck; frontal band and sides of face, whitish, with a black line through the eye; centre of crown, orange-yellow, bordered by a broad black line on either side; ear-coverts, slaty; wings and tail, dark brown, with pale margins, crossed by two white bars; beneath, dull buffish-white." Irides, hazel; bill, legs, and feet, dark brown. Length, about three inches. The female is duller.

Goldcrest.

JOHN DUNCAN.

THE Goldcrest (*Regulus cristatus*, Koch) is a generally and widely distributed resident throughout the British Isles in suitable localities. It is also met with over the whole of Europe, north to the Arctic Circle, and occurs in temperate Asia as far east as the Amoor. The adult has the feathers on crown bright golden orange, paler at sides and bordered with black; forehead, dingy olive; lores, greyish-white; eye-stripe, cheeks, and throat, brownish-yellow; upper parts, olive-green; primaries, greyish-brown, margined with yellowish; secondaries, black on lower parts and tipped with whitish; greater and median wing-coverts, margined with white; under parts, greenish-buff, becoming whiter on abdomen; tail, ash-brown, margined with yellowish; irides, brown; bill, dark brown; legs and feet, brown. Length, about three inches.

Common Nuthatch.

THE Common Nuthatch (*Sitta cæsia*, Wolf) is a very local resident species, being met with principally in England. It is of rare occurrence in Scotland, and, so far, has not been known to occur in Ireland. It also inhabits Central Europe, and is found in Asia Minor and Palestine. The adult has the upper parts slaty-blue; chin, whitish, shading into brownish-buff beneath; flanks, tinged with chestnut; from the base of the bill a black band reaches to each eye, extending behind the eye along the sides of the neck; primaries, greyish-brown; tail-feathers, with the exception of centre ones, black, tipped with grey; outermost rectrices, blackish, tipped with grey and barred with white; irides, hazel; bill, dark horn colour; base beneath, whitish; legs and feet, pale brown. Length, about five inches.

Bearded Titmouse.

JOHN DUNCAN

THE Bearded Titmouse (*Panurus biarmicus*, Linnæus) is a resident species in the British Islands, but is chiefly confined to the vicinity of the Norfolk Broads.   It is also a resident in Europe and Central Asia.   The adult male in summer has the crown grey; moustachial line, black; nape, back, and rump, rufous-brown; scapulars, buffish-white; lesser wing-coverts, greyish-brown, with buff tips; greater wing-coverts, black, broadly margined and tipped with rufous-brown; wings, dark brown; primaries, edged broadly and tipped with white; secondaries, edged with russet; throat, dull white; breast, tinged with flesh colour; sides, reddish-buff; tail, which is graduated, reddish-fawn colour; outer feathers, with white edges; under tail-coverts, black; irides, yellow; bill, yellow; legs and feet, black. Length, about six inches.   The female has no black moustachial line.

Long=tailed Titmouse.

THE Long-tailed Titmouse (*Acredula caudata rosea*, Blyth) is a resident species generally distributed over the British Islands. It also occurs in Western Europe, France, Belgium, and in the Rhine district. The adult male has the head black and white ; above, black and pink ; below, rosy grey ; irides, brown ; bill, legs, and feet, black. Length, about six inches.

Crested Titmouse.

THE Crested Titmouse (*Parus cristatus*, Linnæus) is said to be a "resident only in the forests which clothe the valley of the Spey and the adjacent rivers." It is also found inhabiting Northern and Central Europe. The adult has the feathers on head black, broadly tipped with white; the feathers springing from the crown backwards are considerably elongated and form a conical crest; cheeks, whitish, marked with a few black specks; ear-coverts, black, behind which is a collar of white, margined by a black line reaching from the nape and joining the throat, which, with the chin and upper breast, is also black; upper parts, pale yellowish-brown; primaries and tail, brownish; under parts, white, tinged with ochreous; irides, brown; bill, black; legs and feet, leaden colour. Length, about four inches. The female has a shorter crest, and the black on the throat is less developed.

Marsh Titmouse.

THE Marsh Titmouse (*Parus palustris*, Linnæus, and *Parus palustris Dresseri*, Stejneger) is a resident species in some counties in England and Wales, but in Scotland it becomes very local, and in Ireland it appears to be scarce. It is a common resident throughout Central Europe, but in Southern Europe it becomes scarcer. The adult has the head, nape, and throat, black; upper parts, olive-brown; primaries and tail, ashy-brown; outer margins of feathers, edged with paler brown; sides of head, whitish; under parts, ashy-white, tinged on flanks with buff; irides, brown; bill, black; legs and feet, bluish-grey. Length, about four inches. The female is similar in colour to the male. The young closely resemble the adults, but the colours are not so bright.

Coal Titmouse.

THE Coal Titmouse (*Parus ater*, Linnæus; *Parus ater Britannicus*, Sharpe and Dresser) is a generally distributed, though not very numerous, species throughout the British Isles. It also is a resident in most parts of Europe, as well as North-Western Asia. The adult has the head, nape of the neck, and throat, glossy black; cheeks, sides of the neck, and a central spot in the nape, white; back and scapulars, greenish-brown, tinged with grey; rump, yellowish-grey; wings, greyish-brown, margined with paler brown; the greater and lesser coverts, tipped with dull white, forming two bars across the wings; tail, greyish-brown; breast and abdomen, dull white; flanks, buffish; irides, brown; bill, blackish; legs and feet, bluish-grey. Length, about four inches. The female is similar to the male.

Blue Titmouse.

THE Blue Titmouse (*Parus cœruleus,* Linnæus) is a common resident in the British Isles, and also in Central and Southern Europe. The adult has the crown of the head cobalt blue; forehead, and a line running backward over the eyes encircling the head, white; cheeks, white; streak before and behind the eye, black; nape of the neck, collar, and throat, deep blue-black; back, olive, tinged with yellow; rump, paler; wings and tail, blue; greater coverts and secondaries, tipped with white; under parts, sulphur-yellow; irides, dark brown; bill, bluish; legs and feet, bluish-grey. Length, about four inches. The female is duller in colouration. The young show less blue and more yellow in comparison to their otherwise dingy plumage.

Great Titmouse.

THE Great Titmouse (*Parus major*, Linnæus) is one of the commonest and most widely distributed of the resident titmice found in the British Islands. It is also a resident species in Central and Southern Europe, as well as Eastern Asia, and it occurs in some parts of Northern Africa. The adult has the head and throat glossy black, iridescent with rich blue; ear-coverts and cheeks, white; nape, yellowish-green, showing a whitish spot in the centre; back, yellowish-green; rump and tail, bluish-grey, the latter having the exterior feathers of outer webs white; greater coverts tipped with white, margined externally with grey; primaries, blackish; chest, and centre of abdomen, black; rest of under parts, dull sulphur-yellow; irides, dark brown; bill, black; legs and feet, slate colour. Length, about six inches.

Waxwing.

THE Waxwing (*Ampelis garrulus*, Linnæus) is a nomadic autumn and winter migrant to the British Isles. It is found inhabiting the Northern Nearctic and Palearctic Regions. The adult male has the forehead and crest reddish-chest-nut; from the base of the bill a broad black streak passes through and over the eye; throat, black; upper parts, generally light greyish-brown; primaries, blackish, with the outer web towards the tip white on the outer feather, and yellow on the inner ones; the inner web with broad tips of white; secondaries, blackish-grey, tipped with red sealing-wax-like points, and terminated with a broad bar of yellow; under parts, vinous red; irides, reddish-brown; bill, black; base beneath, pale horn colour; legs and feet, black.   Length, about seven inches.

Red=backed Shrike.

THE Red-backed Shrike (*Lanius collurio*, Linnæus) is a fairly well distributed species in the British Islands during the summer, especially over Wales and Southern and Central England. It is also said to be met with in South Scandinavia, Asia Minor, and Turkestan, moving southward in winter to Africa. The adult male has the crown of the head, nape, upper back, rump, and upper tail-coverts, bright ash-grey; lores, feathers round the eye, and ear-coverts, black; back and scapulars, rich chestnut-brown; wing-coverts, black, edged broadly with chestnut; wings, black; quills, narrowly edged with chestnut; secondaries, with broad margins of the same colour; middle feathers of tail, black; remainder with more or less white at base, and tipped with white; chin, white; lower parts, rose-red; irides, brown; bill, legs, and feet, black. Length, about seven inches. The adult female can be distinguished by having no black about the head.

Woodchat Shrike.

THE Woodchat Shrike (*Lanius rufus*, Brisson) is only an
accidental visitor to the British Isles, mostly during migration.   It is said to be a summer visitant to Central and
Southern Europe, nesting as far east as Persia, and as far
north as the shores of the Baltic.   The adult male has the
forehead, space round eye, ear-coverts, sides of neck and
upper portion of back, black ; lores, white ; crown, nape,
and upper back, rich chestnut ; rump, white ; lower back,
grey; longest upper tail-coverts, grey; tail, black, central
two rectrices white at base and with white tips ; outermost
two tail-feathers nearly all white on outer web ; wings, black ;
basal portion of quills, white, showing a conspicuous spot ;
secondaries, tipped with white ; under parts, white ; irides,
hazel-brown ; bill, legs, and feet, black.   Length, from six
to seven inches.   The female is similar to the male, but the
colouration is duller.

Lesser Grey Shrike.

JOHN DUNCAN.

THE Lesser Grey Shrike (*Lanius minor*, Gmelin) is a rare
abnormal spring and autumn migrant to the British Islands.
It is a summer visitor to Central and Southern Europe,
where it breeds, and winters in Southern Africa. The adult
male has the crown of the head, hinder crown, nape, hind
neck, mantle, scapulars, upper and lower back, pearl-grey;
rump, lighter; tail, black, feathers more or less marked with
white; two outermost ones on each side, all white; frontal
band, feathers round eye, and ear-coverts, black; wings,
black, with a conspicuous white spot; cheeks, white; under
parts, white, tinged with rosy on breast and flanks; irides,
brown; bill, legs, and feet, black. Length, about eight
inches. The female shows less black on the front of the
head.

Great Grey Shrike.

THE Great Grey Shrike (*Lanius excubitor*, Linnæus) is an autumn and winter migrant to the British Islands, and a summer visitor to Northern and Central Europe, spending the winter in Southern Europe. The adult male has the crown and upper parts pearl-grey, getting more ashy towards the tail-coverts, the latter being tipped with white; forehead and eye-stripe, white; wings, black; secondaries, white at bases, and tipped with white; quills, barred with white near their bases, and forming two white spots; lores, and below eye, and auriculars, black; under parts, white; two centre tail-feathers, black, next two, whitish; outermost ones, all white; irides, brown; bill, legs, and feet, black. Length, about nine inches. The adult female is similar to the male, but the colouration is duller.

Pallas's Grey Shrike.

PALLAS's Grey Shrike (*Lanius major*, Pallas) is said to be a fairly frequent autumn migrant to the British Isles, and to inhabit Siberia and Manchuria, south of about lat. 65°. This shrike closely resembles *Lanius excubitor*, but has only one white wing bar, whilst the latter has two. The adult sexes are similar in colour.

Chiffchaff.

THE Chiffchaff (*Phylloscopus rufus*, Bechstein), which is a spring migrant to the British Islands, is also a summer visitant to Northern temperate Europe, spending the winter on the shores of the Mediterranean. The adult in summer has the general colour of the upper parts olive-green; rump, yellowish-green; tail, brown, edged with yellowish-green; wings, brown, with yellowish-green edgings; eye-

6

stripe, pale yellow, shading into white behind ear-coverts; lores, olive; chin and throat, whitish; primaries, brown, edged on the outside webs with olive-green, and narrowly tipped with white; under parts, whitish, tinged with greenish-buff; irides, brown; bill, legs, and feet, blackish-brown. Length, about four inches. The adult female is similar to the male. The young resemble the adults, but underneath they are wholly olive-yellow.

Pallas's Willow Wren.

JOHN DUNCAN.

PALLAS'S Willow Wren (*Phylloscopus proregulus*, Pallas) is an exceedingly rare visitor to the British Isles. An example of this wren was killed on October 31st, 1895, at Cley, in Norfolk. It bears a great resemblance to *Phylloscopus superciliosus*, but can easily be distinguished from the latter by the pale mesial line on the crown, and by the strikingly pale yellow rump.

Blyth's Willow Wren.

JOHN DUNCAN

BLYTH'S Willow Wren (*Phylloscopus viridanus*, Blyth) is a very rare visitor to the British Isles, and is a native of Central Asia, nesting in Turkestan, and wintering in India. A specimen was procured on September 5th, 1895, at North Cotes, on the Lincolnshire coast. This warbler closely resembles *Phylloscopus trochilus*, but shows a conspicuous wing bar.

Yellow=browed Willow Wren.

THE Yellow-browed Willow Wren (*Phylloscopus supercili-osus*, Gmelin) is an exceedingly rare abnormal autumn migrant to the British Isles. It is found inhabiting the North-east Palearctic Region. The adult male in summer is above "greyish-olive, washed with green on rump and upper tail-coverts; wings, brown, crossed by two con-spicuous whitish bars; most of feathers externally margined with yellowish-green; from base of bill to nape a broad greyish-white streak, with a dark one immediately below it and through the eye; beneath, white, washed with yellowish-green, especially on flanks; axillaries, yellow;" irides, dark brown; bill, legs, and feet, brown. Length, about three inches.

Willow Wren.

THE Willow Wren (*Phylloscopus trochilus*, Linnæus) is a common summer visitor to the British Islands from Africa and Persia, moving northwards to the Arctic Regions. The adult in summer has the general colour of the upper parts olive-green, shading into yellower on the rump; wing-coverts, olive, with edgings of olive-green; primaries, brown, tipped narrowly with greyish-white; tail, dull brown; eye-streak and ear-coverts, pale yellow; most of under parts, white, tinged with yellow; breast and flanks, suffused with yellowish-buff; irides, hazel; bill, legs, and feet, brown. Length, about five inches. The female is similar to the male in colouration. The young show more yellow on the under parts than the adults.

Wood Wren.

THE Wood Wren (*Phylloscopus sibilatrix*, Bechstein) is a summer visitor to the British Islands, and also to the temperate parts of Northern Europe.   It wends its way to North Africa to spend the winter.   The adult male in nuptial plumage has the upper parts yellowish-green; a broad streak of sulphur-yellow passes over and behind the eye; wing-coverts, olive-green, margined with paler green; primaries, brown, tipped with whitish, and edged with yellowish-green; tail, greyish-brown; throat and breast, sulphur-yellow; abdomen and under tail-coverts, white; irides, hazel; bill, dark brown, much lighter below; legs and feet, brown.   Length, from four to five inches.   The female is similar in plumage to the male.

Dartford Warbler.

JOHN DUNCAN

THE Dartford Warbler (*Sylvia provincialis*, Gmelin) is a resident species in the British Isles, but is principally confined to the south. It is also found as a resident in the Western and Southern parts of Europe. The adult male in summer has the head deep slate-colour; nape and back, sooty-brown; tail, very dark grey; outer feathers, margined and tipped with whitish; innermost secondaries, wing-coverts, and primary feathers, dark brown, with paler margins; some white streaks on the throat; under parts, chestnut-brown, merging into white on the centre of the abdomen; irides, orange-yellow; bill, very dark brown; base of under mandible, paler; legs, feet, and claws, brown. Length, five inches. The adult female in summer has the under parts light cinnamon-brown, with the exception of the centre of the belly, which is white.

Lesser Whitethroat.

THE Lesser Whitethroat (*Sylvia curruca*, Linnæus), which is a spring migrant to the British Isles, nests throughout most parts of Europe, and winters in Africa. The adult male in summer plumage has the upper parts light slate-grey, suffused more or less with brown on the back, lores, and ear-coverts; eye-stripe, barely discernible; primaries, brown; innermost secondaries, margined broadly with lighter brown; tail, brown; outer tail-feathers on each side, white; breast and flanks, very pale brown; remainder of under parts, white; irides, hazel; bill, blackish, lighter at base beneath; legs, feet, and claws, bluish-grey. Length, about five inches. The female closely resembles the male, but the eye-stripe is absent. The young are similar to the adult female.

Whitethroat.

THE Whitethroat (*Sylvia cinerea*, Bechstein) is a common and widely distributed summer visitor to the British Isles, and is also a summer visitant to most parts of Europe, passing the winter in Southern Africa. The adult male in nuptial plumage has the head and neck ashy-grey; mantle, greyish-brown; upper tail-coverts, ash-grey; tail, dark greyish-brown; outermost tail-feathers, edged with whitish; the next pair tipped broadly with white; wings, brown; secondaries, broadly margined with pale chestnut; throat and abdomen, white, tinged with vinous on breast; flanks, buff; irides, hazel; bill, dark brown; legs and feet, pale brown. Length, about five inches. The adult female shows no vinous tinge on breast, and the rest of the plumage is duller.

Garden Warbler.

THE Garden Warbler (*Sylvia hortensis*, Gmelin), which is a spring and summer visitor to the British Isles, is said to be found everywhere throughout Europe. It winters in Africa. The adult male in summer plumage has the upper parts olive-brown; darker on the wings and tail; eye-streak, pale olive-brown; primaries, dark brown, with lighter margins on wing-coverts and innermost secondaries; underneath, mostly buffish white; centre of abdomen, whiter; breast and flanks, pale olive-brown or buffish-brown; centre of under tail-coverts, buffish-brown; irides, hazel; bill, dark brown; base of lower mandible, paler; legs, feet, and claws, bluish-grey. Length, from five to six inches. The adult female is similar to the male, but a little paler in colouration. The young show broad pale margins on primaries and secondaries.

Blackcap Warbler.

THE Blackcap Warbler (*Sylvia atricapilla*, Linnæus) is a summer visitor to the British Islands, being generally, although somewhat locally, distributed throughout England and Wales. It becomes rarer in Scotland, and in Ireland is very sparingly dispersed. It is a resident in Southern Europe, also in the northern parts of Africa, and in summer extends northwards through the temperate regions of Europe. The adult male in summer has the forehead, crown, and occiput, black; general colour of upper parts, bluish-grey, slightly tinged with olive-brown, most prominently on the margins of primaries and tail-feathers; throat and under parts, greyish-ash; chin and belly, paler; irides, hazel; bill, horn colour; legs and feet, lead colour. Length, about five inches. The adult female has the top of the head reddish-brown; remainder, browner than the male.

Orphean Warbler.

JOHN DUNCAN.

THE Orphean Warbler (*Sylvia orphea*, Temminck) is a very
rare irregular spring migrant to the British Islands.　It
nests throughout the whole of Central and Southern
Europe and Northern Africa.　The adult male has the
general colour of the upper parts slate-grey, with a brown
tinge; wings and tail, brown, broadly margined with ashy;
outermost tail-feathers, white on outer webs; head, lores, and
ear-coverts, blackish; cheeks, throat, and remainder of
under parts, white, tinged on breast and flanks with buffish;
under tail-coverts, buff; irides, pale yellow; bill, blackish,
base beneath, paler; legs and feet, dark brown.　Length,
about six inches.　The adult female is similar to the male,
but the colour of the plumage is duller.　The young resemble
the female.

Subalpine Warbler.

THE Subalpine Warbler (*Sylvia subalpina*, Bonelli) is an exceedingly rare visitor to the British Isles, and is a summer visitor from Northern Africa to the basin of the Mediterranean. The adult male in summer is above, ashy-grey; wings and tail, brown, feathers edged with paler brown; throat and chest, reddish-chestnut, bordered on either side by a white line from base of bill; remainder of under parts chiefly white; irides, brown; eyelids, red; bill and legs, brown. Length, about five inches.

Barred Warbler.

THE Barred Warbler (*Sylvia nisoria*, Bechstein) is a very rare abnormal autumn migrant to the British Islands. It is a summer visitant from North-Eastern and Central Africa to most parts of both Central and Southern Europe. The adult male in summer has the general colour of the upper parts brownish-grey; primaries, wing-coverts, and innermost secondaries, browner, with broad tips of white to the two latter; head and rump, greyish-brown; most feathers above, with the exception of middle of back, have more or less broad tips of white, and a subterminal dark bar; below, white, washed with grey; and barred on chin, throat, breast, sides, and under tail-coverts with brown; irides, light yellow; bill, dark brown, pale at base of under mandible; legs and feet, light slaty-brown. Length, about six inches.

Marsh Warbler.

JOHN DUNCAN

THE Marsh Warbler (*Acrocephalus palustris*, Bechstein), which is said to be the most local of all warblers that visit us for nesting purposes, is a summer migrant and reaches the British shores in May. It has occurred in Somersetshire, Cambridgeshire, and Gloucestershire. This species breeds in Central and Southern Europe, and winters in Africa. The adult closely resembles the reed warbler, but is said to be "more olivaceous-brown above and whiter beneath; and with a more pointed wing; and second primary longer than fourth; legs, brownish flesh colour; irides, dark brown." Length, about five inches. The young are described as being greener on the upper parts than the adults.

Rufous Warbler.

- JOHN DUNCAN

THE Rufous Warbler (*Aëdon galactodes*, Temminck) is an exceedingly rare autumn migrant to the British Islands. It nests in South-Western Europe, and winters in Africa. The adult has the upper parts chestnut; wing-coverts and primaries, brown, margined on outside web with creamy-white; superciliary stripe, creamy-white; tail, bright chestnut, banded with black at terminal end, and followed by tips of white to all the feathers excepting central pair; beneath, creamy-white; irides, dark hazel; bill, above, brown; below, paler at base; legs and feet, light brown. Length, about six inches. The female resembles the male. The young are similar to the adults, but are said to have the primaries and wing-coverts broadly margined with light fulvous.

Reed Warbler.

JOHN DUNCAN

THE Reed Warbler (*Acrocephalus arundinaceus*, Brisson)
is a summer visitor to our islands, its principal headquarters
being in the Southern, Midland, and Eastern Counties. Its
range outside the British Islands is the same as that of the
marsh warbler. The adult in summer garb has the general
colour of the upper parts olive-brown, tinged with rufous,
especially on the rump and upper tail-coverts; wings and
tail, brown, edged with paler brown; under parts, white,
shading into buff on the sides, thighs, and under tail-
coverts; a pale buff streak passes over each eye; irides,
brown; bill, pale brown, base of lower mandible paler;
legs, slaty-brown. Length, about five inches. The adult
female closely resembles the male. The young are similar
in colouration, but show more buff underneath.

Sedge Warbler.

THE Sedge Warbler (*Acrocephalus phragmitis*, Bechstein) is a summer visitor to the British Isles, and is commonly distributed in all suitable localities. It is said to breed throughout most parts of Europe, and occurs as far eastward as the valley of the Yenisei, and a short distance north of the Arctic Circle. The adult in summer has the upper parts russet-brown, with dusky-brown centres to the feathers; eye-stripe, buffish-white; rump, yellowish-rufous; crown, streaked with brown; wings and tail, brown, margined with paler brown; beneath, buffish-white, tinged on flanks with yellowish-brown; irides, hazel; bill, dark brown, paler below; legs, feet, and claws, pale brown. Length,

7

five inches.   The female shows less rufous on the rump, and the general colour of the plumage is duller than the male.

Great Reed Warbler.

THE Great Reed Warbler (*Acrocephalus turdoides*, Meyer) is a very rare irregular spring and autumn migrant to the British Isles.   It breeds in both Central and Southern Europe, and in winter is a resident in Africa.   The adult has the general colour above, pale brown, tinged on wings, rump, and tail with rufous-brown ; superciliary stripe, very indistinct ; chin, throat, and centre of belly, nearly white ; rest of under parts, light rufous-brown ; irides, brown ; bill, dark brown ; underneath, light brown ; legs, light horn colour.   Length, about eight inches.   The young show more red on the upper parts, and more fawn colour below.

Aquatic Warbler.

JOHN DUNCAN.

THE Aquatic Warbler (*Acrocephalus aquaticus*, Gmelin) is a very rare abnormal autumn migrant to the British Islands, breeding in Central and Southern Europe from the Atlantic to the Urals. It is supposed to winter in Africa. The adult bears a close resemblance to the sedge warbler (*Acrocephalus phragmitis*), but can readily be distinguished from that bird by its having only two broad, distinct, and conspicuous dark stripes on the crown, and being less in size.

Melodious Warbler.

JOHN DUNCAN.

THE Melodious Warbler (*Hypolais polyglotta*, Viell) has
one British record, and is found inhabiting South-Western
Europe and North-Western Africa.   The adult male in
summer is, above, greyish-olive, darker on head; superciliary streak, yellowish; wings and tail, brown, margined
with lighter; beneath, pale yellow; sides of neck and
flanks, tinged with pale brownish-olive; irides, brown;
bill on upper part, brown; underneath, yellowish at base;
legs and feet, greyish-brown.   Length, about four inches.

Grasshopper Warbler.

THE Grasshopper Warbler (*Locustella locustella*, Latham) is a summer visitor to the British Isles, and found nesting in most parts of England and Wales; it, however, becomes rarer northwards of Northumberland and Durham. It is distributed over the greater part of Europe, and is found as far north as St. Petersburg, wintering in Northern Africa, and it is said occasionally to winter in Southern Europe. The adult male has the upper parts olive-brown, with dark centres to the feathers; primaries and tail, brown, with faint bars on the latter; beneath, pale brown; neck and breast, spotted with darker brown; dark centres on under tail-coverts; irides, brown; bill, legs, and feet, brown. Length, about five inches. The adult female closely resembles the male in colouration.

Savi's Warbler.

Savi's Warbler (*Locustella luscinioides*, Savi) at one time was a regular summer visitant to the Fen districts of England; but, owing to the drainage of the marshes which it used to frequent, it is now in all probability extinct in the British Islands. Savi's warbler is locally distributed through Central and Southern Europe. The adult is uniform russet-brown above; eye-stripe, indistinct; under parts, pale buffish-brown, shading into nearly white on the throat and centre of abdomen; under tail-coverts, pale chestnut, with obscure pale tips; irides, hazel; bill, dark brown above, horn colour below; legs, feet, and claws, light brown. Length, about five inches. The adult sexes are said to be alike.

Icterine Warbler.

THE Icterine Warbler (*Hypolais hypolais*, Linnæus) is a rare accidental spring and autumn migrant to the British Isles, breeding in Northern and Central Europe. On migration, it passes through Southern Europe, spending the winter in Africa. The adult in summer is, above, olive-green; eye-stripe and axillaries, greenish-yellow; wings and tail, brown, with whitish margins; under parts, yellow, tinged with green; irides, brown; bill on upper mandible, brown; beneath, horn colour; legs and feet, greyish. Length, about five inches. The adult female is similar to the male.

Song Thrush.

THE Song Thrush (*Turdus musicus*, Linnæus) is generally distributed throughout the British Isles. It is more or less migratory in its habits, leaving in the late autumn and returning in early spring. The adult is dark olive-brown on the upper parts; eye-stripe, buff; quills and wing-coverts, margined with deep yellowish-brown, and there are buff tips to some of the feathers; under parts, buff, shading into white on belly; sides of throat, chest, and remainder of body, profusely spotted and blotched with dark brown; irides, legs, and feet, pale brown; bill, dark brown. Length, about eight inches.

Missel Thrush.

THE Missel Thrush (*Turdus viscivorus*, Linnæus) is a resident and widely distributed species throughout the wooded districts of the British Islands. It is also found as a resident in most parts of temperate Europe. In those countries where the winters are severe, it migrates to South Europe and North Africa. The adult has the head and upper parts greyish-brown, tinged with rufous; rump, tinged with yellowish; tail, ash-grey; outer feathers, tipped with dull white; inner webs of outermost feathers, also whitish; under parts, buffish-white, boldly spotted with large dark brown fan-shaped spots, becoming smaller and more triangular on the throat; irides, dark brown; bill, dark brown, base beneath paler; legs, pale brown. Length, about eleven inches. The female closely resembles the male.

Redwing.

J.HN DUNCAN.

THE Redwing (*Turdus iliacus*, Linnæus) is a common autumn and winter visitor to the British Isles, and breeds throughout most parts of the Northern Palearctic Region from Norway to the valley of the Yenisei. The adult can easily be distinguished from the song thrush (*Turdus musicus*) by its broad and well-defined white eye-stripe, and by the rich chestnut on the sides of the body, under wing-coverts, and axillaries.

Fieldfare.

JOHN DUNCAN.

THE Fieldfare (*Turdus pilaris*, Linnæus) is a common autumn migrant to the British Isles; it also occurs during winter in Southern and Western Europe, and during the nesting time inhabits the northern pine regions of Europe and Asia. The adult male in summer has the head, neck, and rump grey, the colouration of the rump being brighter; feathers on crown, marked with blackish centres; middle of back, wing-coverts, and scapulars, chestnut; wings and tail, blackish-brown, with narrow pale margins to the former; wings, lighter in colour than the tail; eye-stripe, whitish; cheeks, throat, and breast, ochreous-buff, streaked with black; flanks, marked boldly with blackish-brown; centre of abdomen, white; bill, yellow; irides, brown; legs and feet, blackish. Length, about ten inches. The female closely resembles the male.

Blackbird.

JOHN DUNCAN.

THE Blackbird (*Merula merula*, Linnæus) is a widely distributed resident species in the British Islands. It is also
a resident in most parts of Europe, the Azores, Northern
Africa, and Palestine. In summer it visits those regions only
which are immediately south of the Arctic Circle. The
adult male has the whole of the plumage glossy black, with
the exception of the primaries, which are a shade browner;
irides, dark brown; bill, orange-yellow; legs and feet, dark
brown. Length, from ten to eleven inches. The adult
female is easily recognised by its umber-brown and rufous
colouration. The young females are, above, blackish-brown,
with most of the feathers streaked with pale rufous; under
parts, light rufous-brown.

Ring Ouzel.

THE Ring Ouzel (*Merula torquata*, Linnæus) is a summer visitor to the British Isles. It is also found in the mountainous regions of temperate Europe, and in winter is a resident in Northern and Central Africa, also in Asia Minor. The adult male in summer is blackish-brown on the upper and under parts of the body, the feathers margined with paler more or less; on the upper part of the breast there is a broad gorget of white; axillaries, greyish-brown; irides, dark brown; bill, yellow; legs and feet, brown. Length, between ten and eleven inches. The female is browner than the male; the gorget is duller, being tinged with brownish. The young have the upper and under parts barred with brown and black.

Black=throated Ouzel.

JOHN DUNCAN.

THE Black-throated Ouzel (*Merula atrigularis*, Temminck) is an exceedingly rare autumn migrant to the British Islands, and breeds in Siberia, as far south in Asia as the Himalayas and Turkestan, and in winter south to Northern India and Persia.   The adult male has the upper parts olive-brown; throat and breast, black; abdomen, white; sides and flanks, greyish-brown; tail, dark brown; axillaries, rich chestnut; irides, dark brown; bill, dark brown on upper mandible, lighter below; legs and feet, light brown. Length, about nine inches.   The adult female lacks the black on the throat and breast, the feathers having dark centres; lower throat, uniform buffish-white.   The young males are said to be similar to the adult females.

White's Ground Thrush.

White's Ground Thrush (*Geocichla varia*, Pallas) is a very rare abnormal autumn migrant to the British Islands from Southern Central Asia.   The adult (which cannot well be confounded with any other of our British thrushes) has the upper parts of plumage olive-brown, with black tips to the feathers; under parts white, tinged with buff, and marked with black crescent-shaped spots; the tail consists of fourteen feathers.   Length, about twelve inches.

Siberian Ground Thrush.

JOHN DUNCAN.

THE Siberian Ground Thrush (*Geocichla sibrica*, Pallas),
which is a very rare irregular autumn migrant to the
British Isles, is a straggler from Northern Asia. The
adult male has the general colour of the upper parts deep
slate-grey, margined with paler; central tail-feathers, dark
slate-grey; rest, blackish, the two outer ones tipped with
white; eye-stripe, white and broad; under parts, greyish,
changing into white on centre of belly; under tail-coverts
have white tips; irides, deep brown; bill, blackish; legs

and feet, pale brown. Length, about nine inches. The adult female is brownish on the upper plumage ; buffish on breast, shading into white on belly; flanks, tinged with brown.

Rock=Thrush.

THE Rock-Thrush (*Monticola saxatilis*, Linnæus), which is an exceedingly rare spring migrant to the British Islands, is a summer visitant to certain portions of Central Europe and Siberia. The adult male has the scapulars blackish, tinged with blue ; middle of back, white ; head and neck, slate-blue ; wing-feathers, brown ; upper tail-coverts and tail, orange-chestnut ; the two central tail-feathers darker ; throat, greyish-blue ; remainder of under parts, orange-chestnut ; irides, hazel ; bill, black ; legs and feet, warm brown. Length, about eight inches. The adult female has mottlings of brown above and below ; throat, white, mottled with brown ; tail, lighter than in the male.

8

Robin.

JOHN DUNCAN.

THE Robin (*Erithacus rubecula*, Linnæus) is a well-known
resident species in the British Islands.   It is also a resident
species in the temperate parts of Europe and Northern
Africa.   The adult male has the head and upper parts of
body olive-brown ; forehead, throat, and upper portion of
breast, rich orange-chestnut; sides of neck, pale bluish-grey;
centre of belly, white ; flanks and under tail-coverts, buffish-
brown; irides, black ; bill, black ; legs and feet, pale brown.
Length, about five inches.   The female closely resembles
the male.   The young in first plumage have the upper and
under parts yellowish-brown, with dark tips to the feathers.

Arctic Blue=throated Robin.

THE Arctic Blue-throated Robin (*Erithacus suecica*, Brehm) is an abnormal spring and autumn migrant to the British Isles. There are, says Seebohm, three forms of this bird:—First, the Northern or Arctic form, with the spot in the centre of the throat red; secondly, the South European form, with the spot in the centre of the throat pure and silky white; and thirdly, the form of which the throat is uniform blue. Length, about five inches.

Nightingale.

THE Nightingale (*Erithacus luscinia*, Linnæus) is another summer visitor to the British Isles, and is also met with in Central and Southern Europe. It makes its way into Africa to spend the winter. The adult has the upper parts russet-brown, shading into rusty-red on upper tail-coverts and tail; lores, buffish-white; beneath, buffish-white, shading into greyish-white on breast and flanks; axillaries and under wing-coverts, brownish-white; irides, hazel; bill, brown above, paler beneath; legs and feet, brown. Length, about six inches. The female closely resembles the male. The young in first plumage have pale centres to most of the feathers on upper parts; feathers of under parts have dark margins.

Redstart.

THE Redstart (*Ruticilla phœnicurus*, Linnæus) is a summer visitant to the British Islands. It is also found breeding in Central and Northern Europe, south of the Arctic Circle, and in autumn migrates through Southern Europe to Northern Africa. The adult male has the forehead and eye-streak white; head and back, slate-grey; wings, brown, edged with paler brown; rump, upper tail-coverts, and tail, rich chestnut, with the exception of the two central feathers of latter, which are brown margined with rufous; breast and flanks, rich chestnut, shading into paler on abdomen; a narrow band at base of upper mandible, sides of face, sides of neck, and throat, rich black; irides, dark brown; bill, black; legs and feet, blackish. Length, about five inches. The female is in general colouration brown, with tail and vent dull chestnut.

Black Redstart.

THE Black Redstart (*Ruticilla tithys*, Scopoli) is a fairly regular winter visitor to the British Isles, mostly along the South Coast, including Cornwall, where it is frequently met with.   It is a resident in some parts south of the Alps, but in winter the majority migrate to Northern Africa.   The adult male has the general colouration of the plumage slate-grey; wings, brown, with white margins on the out-side webs of the secondaries; rump, upper and under tail-coverts, bright chestnut; central feathers of the tail, brown, externally margined with bright chestnut; frontal band and lores, black; chin, throat, cheeks, and breast, black; belly and flanks, buffish; irides, brown; bill, black; legs and feet, blackish.   Length, about six inches.

Wheatear.

THE Wheatear (*Saxicola œnanthe*, Linnæus) is a spring
visitor to the British Isles. In summer it is said to be
a "visitant to Central and Northern Europe, extending to
Iceland and Greenland, and eastwards throughout Northern
Siberia. In winter it ranges from Western and Northern
Africa to Persia and Northern India." The adult male in
summer has the upper parts bluish-grey; wings, brownish-
black; lower portion of neck and breast, warm buff; eye-
stripe, forehead, and rump, white; sides of face, black;
abdomen and vent, white; the two centre tail-feathers,
blackish-brown, nearly to the base; remainder white, tipped
broadly with black; irides, dark brown; bill, legs, and feet,
black. Length, between five and six inches. The female
has the plumage sandy-brown, deepest above; the wings
and tail resemble those of the male in colouration. The
young are like the female, but show spots on upper and
under parts.

Isabelline Wheatear.

JOHN DUNCAN.

THE Isabelline Wheatear (*Saxicola Isabellina*, Ruppell), which is an extremely rare autumn migrant to the British Islands, is a resident in North-East Africa (from Abyssinia to Egypt) and Palestine. It nests also in South-East Russia and temperate Asia, and in winter migrates southward. Howard Saunders says that this wheatear may easily be mistaken for the female of the common wheatear, but the broader lining to the quills will always distinguish it. According to Bowdler Sharpe, it is also longer in the tarsus. Length, about six inches.

Black-throated Wheatear.

JOHN DUNCAN.

THE Black-throated Wheatear (*Saxicola stapazina*, Vieillot) is a very rare abnormal migrant to the British Isles; it also occurs in the South of France and Spain, and is a resident in North-West Africa. The male is described as having the head and back sandy-rufous; rump, white; upper and under parts of wings, black. Length, about five inches.

Desert Wheatear.

JOHN DUNCAN

THE Desert Wheatear (*Saxicola deserti*, Temminck) is a
rare autumn migrant to the British Islands. During summer it occurs in Algeria, Egypt, Arabia, and Palestine, and
in winter it ranges through Abyssinia to North-West India.
The adult male in summer has the general colour of the
upper parts buff, changing into white on rump and upper
tail-coverts; tail, black, with white bases; superciliary
stripe, whitish; cheeks, throat, and sides of neck, black;
wings, blackish-brown, with pale tips; under parts, white,
with a tinge of sandy-rufous on chest and flanks; irides,
dark brown; bill, legs, and feet, black. Length, from
five to six inches. In this species the feet are very small.
The adult female in summer has no black on the throat.

Whinchat.

THE Whinchat (*Pratincola rubetra*, Linnæus) is a fairly well distributed summer visitant to the British Islands. It spends the summer throughout temperate Europe, and in winter is a resident in Northern Africa. The adult male has the crown of the head, back, and wing-coverts, brownish-black; the feathers margined with ochreous-yellow; spot on bastard wing, white; a white line passes from the base of the bill over the eyes reaching to the nape of the neck; ear-coverts and cheeks, brownish-black; chin and a streak along the sides of the neck, white; rump, yellowish-brown, with streaks of blackish-brown; tail, dark brown, with bases of all outermost feathers white; centre of throat and breast, fawn colour, shading into pale buff on belly; irides, brown; bill, legs, and feet, black. Length, about five inches.

Stonechat.

THE Stonechat (*Pratincola rubicola*, Linnæus) is a somewhat generally dispersed resident, and partially migratory, species in the British Islands. It is also found inhabiting the temperate parts of Europe west of the Volga. During the winter it is a resident in Africa. The adult male in nuptial plumage has the sides of the neck and upper parts of wings white; rump, white, with dark centres and rufous margins; tail, black, with brown margins; back, black, feathers margined with yellowish-brown; wings, brownish-black; sides of face and throat, black; under parts, rufous-brown; abdomen, yellowish-white; irides, brown; bill, legs, and feet, black. Length, about five inches. The female is brown in the upper part, streaked with darker brown; rump, brown, tinged with red; the throat shows black mottlings.

Hedge Accentor.

THE Hedge Accentor (*Accentor modularis*, Linnæus) is a very generally distributed resident in the British Isles.   It is also a resident species throughout most portions of Europe, and is met with in Scandinavia as far as the limits of forest growth.   The adult male has the top of the head and nape greyish-brown, streaked with brown; sides of the neck, throat, and breast, bluish-grey; back and scapulars, blackish, with broad margins of reddish-brown; primaries, dark brown, edged and tipped with a lighter shade of brown; tail, dull brown, margined with paler; rump, yellowish; chin, whitish-grey; centre of belly, greyish-white; flanks, pale brown, streaked with darker brown; irides, hazel; bill, legs, and feet, pale brown.   Length, about five inches.   The female closely resembles the male.

Alpine Accentor.

THE Alpine Accentor (*Accentor alpinus*, Gmelin), which is a rare straggler during autumn to the British Isles, is a resident in the mountain ranges of Southern Europe.   The adult male has the upper parts dark brown, edged with light brown; head and neck, grey, with darker streaks; wings and tail, blackish-brown, margined with chestnut-brown, and tipped with white; wing-coverts, prominently tipped with white; throat, white, with black spots; breast and under tail-coverts, dullish grey; flanks, variegated with chestnut and light buffish; irides, brown; bill, blackish-brown; base below, yellow; legs and feet, warm brown. Length, about seven inches.

Dipper.

THE Dipper (*Cinclus aquaticus*, Bechstein) is a resident in the British Islands. It is also found, subject to some slight modification of colour, in many of the mountain regions of Central Europe. The adult has the head and back of neck umber-brown; remainder of upper parts, dark slaty-grey, with paler margins to feathers on the back; throat, sides of neck, and upper breast, white; lower part of breast and abdomen, chestnut-brown, shading into brownish-black towards the vent; under tail-coverts, blackish, tinged with grey; flanks, greyish; irides, brown; bill, black; legs and feet, brown. Length, from six to seven inches. The female is similar in colour to the male. The young are greyish-brown on upper parts, and chestnut-brown on under parts absent.

Black=bellied Dipper.

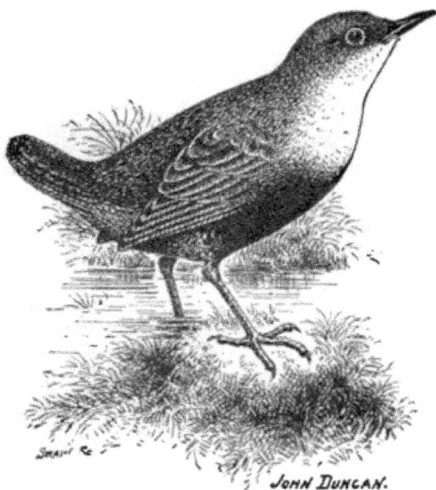

JOHN DUNCAN.

THE Black-bellied Dipper (*Cinclus aquaticus melanogaster*, Brehm) is a rare abnormal autumn migrant to the British Islands; it also occurs in Scandinavia, Northern Germany, and Northern Russia. The adult male is similar in colouration to the common dipper, *Cinclus aquaticus*, with the exception that the breast is black or nearly black, and not chestnut-brown as in the latter bird. Indeed, the black-bellied dipper is considered merely a form of the common dipper.

Common Wren.

THE Common Wren (*Troglodytes parvulus*, Koch) is a widely and generally distributed resident species in the British Islands. It is also a common resident in most parts of temperate Europe, and occurs in Northern Africa and Western Asia. The adult has the general colour above, dark reddish-brown, barred with dark brown; eye-stripe, whitish; rump, rufous-brown; wings and tail, rufous-brown; under parts, buffish-white; abdomen and flanks, tinged with brownish, and transversely barred with brown; irides, brown; bill, brown; legs and feet, pale brown. Length, about four inches. The female resembles the male. The young are similar in ground colour to the female, but the plumage is less barred.

9

St. Kilda Wren.

JOHN DUNCAN.,

THE St. Kilda Wren (*Troglodytes hirtensis*, Seebohm) is a resident species on the islands of St. Kilda, Scotland. To Mr. Charles Dixon, the well-known naturalist, is due the honour of having brought from St. Kilda the specimens originally described by the late Henry Seebohm. The adult male closely resembles the common wren, but is superior in size, and the feet are much stronger and larger.

Spotted Flycatcher.

THE Spotted Flycatcher (*Musicapa grisola*, Linnæus) is a late summer migrant to the British Islands. It is found nesting almost everywhere in Europe, and winters in Africa. The adult male in breeding plumage has the upper parts of the body pale hair-brown, the crown of the head having darker centres to the feathers; wings and tail, darker brown; throat and centre of abdomen, white; sides of neck, breast, and flanks, streaked with hair-brown; irides, brown; bill, dark brown; legs and feet, black. Length, about five inches. The female is similar to the male. The young have the feathers of the upper parts marked in centre with buff.

Pied Flycatcher.

THE Pied Flycatcher (*Musicapa atricapilla*, Linnæus) is a spring visitor to the British Islands. It is found breeding in Europe south of the Arctic Circle, and spends the winter in Central Africa. The adult male in summer has the head and upper parts of body black; rump, greyish; forehead, sides of neck, and under parts, pure white; wings, brown; innermost secondaries, edged broadly with white; central coverts, white; tail, blackish; outermost feathers, partially white; irides, brown; bill, legs, and feet, black. Length, about five inches. The female is much more dingy in colouration than the male. The young in nestling plumage are spotted on the upper parts with buff; under parts, spotted irregularly with blackish-brown.

Red=breasted Flycatcher.

THE Red-breasted Flycatcher (*Musicapa parva*, Bechstein) is an occasional autumn migrant to the British Isles, occurring also in both Central and South-Eastern Europe, and as far to the east as India, wintering in Northern Africa. The adult male in summer has the general coloration of the upper parts of plumage olive-brown; crown, nape, and cheeks, bluish-grey, paler on latter; two middle tail-feathers, blackish-brown; base of outer ones, white; terminal end, broadly tipped with blackish-brown; throat and breast, deep orange with a reddish tinge and fringed with greyish; remainder of under parts, white, washed on flanks and under tail-coverts with creamy-buff; irides, hazel; bill, brown, lighter at base; legs and feet, brown. Length, from four to five inches. The female in general resembles the male, but the throat is buff.

Barn Swallow.

JOHN DUNCAN

THE Barn Swallow (*Hirundo rustica*, Linnæus) is a well-known summer visitant to the British Isles. It is also a common visitor to Europe during summer, and passes the winter in Ethiopia and India. The adult male has the fore-head and throat rich chestnut; sides of neck, back, wings, and a broad band across breast, deep blue, with purplish reflections; wings and tail, brown, outer webs glossed with bluish-green; all tail-feathers, with the exception of centre ones, have a mesial spot of white; abdomen, light buff; under tail-coverts, pale chestnut; irides, brown; bill, legs,

and feet, black.   Length, about seven inches.   The female
shows less chestnut on the forehead, and the two outer tail-
feathers are shorter than in the male.

House Martin.

THE House Martin (*Chelidon urbica*, Linnæus) is a summer
migrant to the British Isles.   It nests throughout most
parts of Europe and Northern Africa, and is found in
winter south of Abyssinia.   The adult has the head, neck,
and back bright glossy bluish-black ; rump, white ; wings
and tail, dullish black ; under parts, white ; irides, dark
hazel ; bill, black ; legs and feet covered with white feathers.

Length, about five inches. The female closely resembles the male in colouration. The young are sooty-brown on the upper parts; rump and under parts, dullish white.

Sand Martin.

THE Sand Martin (*Cotyle riparia*, Linnæus) is a spring visitant to the British Islands. It is also a visitor to every part of the Palearctic and Nearctic Regions; and is found during winter in India, Africa, and South America. The adult male has the upper parts, cheeks, and a band across chest, hair-brown; primaries and tail-feathers, darker brown; throat and abdomen, whitish; flanks, brownish; irides, hazel; bill, black; legs and feet, dark brown. Length, about five inches. The female is the same in plumage as the male. The young have the feathers above tipped with whitish; under parts, washed with buff.

Purple Martin.

JOHN DUNCAN.

THE Purple Martin (*Progne purpurea*, Linnæus) is a rare abnormal autumn migrant to the British Isles, and a summer visitant to the United States and Canada, moving northwards above the Arctic Circle. It spends the winter in Mexico, and is said to have been met with in the Bermudas. The adult male has the general colouration of the plumage steel-blue, with purple reflections; wings and tail, black, tinged with blue; irides, brown; bill, black; legs and feet, brown. The female closely resembles the male, but the colouration is duller, and the under parts are of a greyish-brown. The young birds are described as being similar to the adult females.

Wryneck.

THE Wryneck (*Yunx torquilla*, Linnæus) is a spring visitor to the British Isles; it is also found inhabiting the Palearctic Region, with the exception of the extreme north, wintering in Northern Africa, India, and China. The adult has the ground colour above, yellowish-grey, beautifully mottled and vermiculated with blackish-brown; wings, brown, having dull chestnut bars on the outer webs; under parts, buff, with a dark brown subterminal spot or bar on each feather; tail, ashy-brown, vermiculated and barred with black; irides, pale brown; bill, legs, and feet, brown. Length, between six and seven inches. The female is similar to the male. The young are duller above, and conspicuously marked underneath with black.

Green Woodpecker.

THE Green Woodpecker (*Gecinus viridis*, Linnæus) is a
resident species in the British Islands, and also met with
in the Western Palearctic Region; northwards to South
Scandinavia, and ranging southwards to the Mediterranean
and the Pyrenees. The adult male has the crown of the
head and nape scarlet, with grey bases; lores, cheeks, ear-
coverts, and feathers round eye, black; moustachial band,
black, washed with crimson; mantle, dull light green;
rump, yellow; wings and tail, dark greenish-brown, the
latter obscurely barred with darker; primaries, dusky,
barred on outer web with white; under parts, yellowish-
grey, tinged with greyish-green; irides, whitish; bill, dark
grey, blackish at tip and paler at base beneath; legs and
feet, dark grey. Length, between eleven and twelve inches.
The female shows no scarlet on the moustachial lines.

Lesser Spotted Woodpecker.

THE Lesser Spotted Woodpecker (*Picus minor*, Linnæus) is a resident species in the British Islands; it is also a resident in most parts of the Western Palearctic Region, but is said not to occur in Egypt, Palestine, Iceland, or the Faroes. The adult male has the upper parts black, with broad bars of white; central tail-feathers, black, remainder, black, with white bars; forehead, buff; crown, crimson; eye-streak, occiput, and nape, black; moustache, black; cheeks and sides of face, white; chin, throat, and breast, dull white, tinged with brown on sides; flanks, streaked with black; irides, reddish-brown; bill, legs, and feet, dark grey. Length, about five inches. The female shows no red on the crown. The young are similar to adults, but the markings are not so distinct.

Great Spotted Woodpecker.

THE Great Spotted Woodpecker (*Picus major*, Linnæus) is a resident in the British Isles, and is also pretty generally distributed throughout the whole of the European Continent. The adult male has the general plumage of the upper parts black; forehead, white; cheeks and ear-coverts also white; nape, crimson; a broad black band runs from base of bill to nape; another band of black traverses sides of chest; scapulars, white; primaries and secondaries, barred exteriorly with white; on each side of the back part of the neck is a white patch; throat, breast, and abdomen, whitish; vent and under tail-coverts, crimson; rump and central tail-feathers, black; outer ones with alternate bars of black and white; irides, pale red; bill, legs, and feet, slaty-black. Length, about nine inches. The female shows no red on the head.

Cuckoo.

THE Cuckoo (*Cuculus canorus*, Linnæus) is a well-known spring migrant to the British Isles. It is also a summer visitor to Europe, as well as most portions of Asia. In the autumn it migrates to Central Africa and Southern India. The adult male has the head, neck behind, nape and back, dark ash colour; wing-coverts, darker ash colour; primaries, dusky, with oval white spots on the inner webs; tail, blackish, with some of the webs spotted with white, and all the tail-feathers tipped with white; throat and upper breast, pale ash colour; under parts, dullish white, with transverse black bars; irides and feet, yellow; bill, dusky; base beneath, yellowish. Length, about thirteen inches. The female is similar to male. The young on upper parts are warm brown with darker transverse markings.

Great Spotted Cuckoo.

THE Great Spotted Cuckoo (*Coccystes glandarius*, Linnæus) is a very rare straggling spring and autumn migrant to the British Isles, and is also found inhabiting South·West Europe and North Africa.   The adult male has the upper parts greyish-brown, most of the feathers being tipped with white; crest and crown, grey; sides of face, dark grey; tail-feathers, blackish-brown, with white tips to all but the central ones; neck, creamy-white; under parts, dullish white; irides, brown; bill, dark brown; base beneath, yellowish; legs and feet, slate-grey.   Length, from sixteen to seventeen inches.   The adult female has a smaller crest, and the ground colour of the primaries is rufous.   The young are similar to the adult females, but show little crest; and the under parts are washed with chestnut.

Black-billed Cuckoo.

THE Black-billed Cuckoo (*Coccyzus erythrophthalmus*, Wilson) is an exceedingly rare abnormal autumn migrant to the British Isles.   It is a North American species, ranging from Canada to Brazil.   In winter it migrates from the northern portions of Canada.   This cuckoo can be distinguished from *Coccyzus Americanus*, being slightly less in size, and showing very little chestnut on the quills.   The bill is nearly black ; the orbits, dull scarlet ; and the white tips at the end of the tail are much less conspicuous.

Yellow=billed Cuckoo.

JOHN DUNCAN.

THE Yellow-billed Cuckoo (*Coccyzus Americanus*, Linnæus)
is a very rare abnormal autumn migrant to the British
Islands.   It inhabits America, from Canada south to
Brazil, and in winter migrates from the northern portions.
The general colouration of the upper parts is buffish-brown,
glossed with greenish; two central tail-feathers, buffish-
l rown; remainder, nearly black, tipped broadly with white;
quills, tinged with pale yellowish-chestnut; under parts,

10

white; orbits, whitish; irides, dark brown; upper mandible, blackish, with a yellow base; under mandible, yellow, with a nearly black tip; legs and feet, grey.

Common Swift.

THE Common Swift (*Cypselus apus*, Linnæus) is a late summer migrant to the British Isles, and is also a visitor in summer to most parts of the Palearctic Region.   It spends the winter in Africa. The adult has the general colour of the upper parts of plumage blackish-brown, with bronze-like reflections; chin and throat, dullish white; under parts, blackish-brown; irides, dark brown; bill and feet, black; tarsi, covered with small feathers.   Length,

about eight inches.   The female is same in colour as the
male.   The young are browner, and dullish white on the
chin and throat.

White=bellied Swift.

THE White-bellied Swift (*Cypselus melba*, Linnæus) is a rare
spring and autumn migrant to the British Isles.   In summer
it visits Central and Southern Europe, also Northern
Africa, spending the winter in Africa.   The adult has the
upper parts greyish-brown ; forehead, paler, with a blackish
patch in front of the eye; throat and abdomen, white; under
tail-coverts, greyish-brown ; irides, nearly black ; bill and
feet, black.   Length, about eight inches.   The young are
similar to the adults, but have whitish margins to most of
the upper feathers.

Needle-tailed Swift.

John Duncan.

THE Needle-tailed Swift (*Chætura caudacuta*, Latham) is an exceedingly rare abnormal autumn migrant to the British Isles, and is found inhabiting Eastern Siberia, China, and the Himalayas. It occurs during winter in Australia. The forehead of the adult is dull white; crown, nape, back, and sides of head, dusky black; wing-coverts and secondaries, bottle-green; inner webs of inner secondaries, mostly white; quills, blackish; tail-feathers, bottle-green, with projecting spinous shafts; throat, breast, and under tail-coverts, white; abdomen, sooty brown; flanks, white, marked with glossy blue-black; bill, black; legs and feet, dark brown. Length, about eight inches.

Common Nightjar.

THE Common Nightjar (*Caprimulgus Europæus*, Linnæus) is another of our latest summer visitors to the British Isles. It is also found during summer in the Western Palearctic Region, and in winter goes to Africa and India. The adult male has the general colouration of the plumage of an ashy-grey, spotted and streaked with dark brown, yellowish-brown, and reddish-buff; wings, dark brown, tinged slightly with chestnut, and showing three bars of yellowish-brown; the primaries have a large white patch near the centre of the three first quill feathers of each wing; tail, marked with zigzag bars of black, yellowish-brown, and grey; two outer rectrices on each side, tipped with white; on each side of the throat is a white spot; under parts, light orange-buff, barred with dusky; irides and bill, black; legs and feet, brownish-red Length, ten inches. The female is duller; white spots on the tail and wings very indistinct.

Red=necked Nightjar.

THE Red-necked Nightjar (*Caprimulgus ruficollis*, Linnæus)
is an exceptionally rare migrant to the British Isles.   It is
also met with in South-West Europe and North-West Africa.
This nightjar bears a strong resemblance to the common
nightjar, but it can be distinguished from the latter by the
colour of the nape, which is reddish-buff.   The white spots
on the neck, too, are larger, and have more creamy colour
on the wing-coverts and under parts.

Egyptian Nightjar.

THE Egyptian Nightjar (*Caprimulgus Ægyptius*, Lichten-
stein) is a very rare abnormal spring migrant to the British
Islands.   In summer it is found in Turkestan, Baluchistan,
Egypt, and Nubia; and is supposed to winter farther south.
The ground colour of the whole of the adult's plumage is

isabelline, or sandy-grey, dusted and otherwise marked with white, grey, and brown. The general colouration is much lighter than that of *Caprimulgus Europæus*. Length, about ten inches.

Bee-eater.

THE Bee-eater (*Merops apiaster*, Linnæus) is a rare irregular spring and autumn migrant to the British Isles, and is a visitor in summer to countries adjacent to the Mediterranean, in winter migrating to Africa. The adult has the top of the head, nape, back, and wing-coverts, rich chestnut brown, changing into light greenish-blue on upper tail-coverts; tail, greenish or bluish; two centre pair of rectrices elongated and tipped with black; lores and auriculars, black; chin and throat, deep yellow, with a band of black underneath; scapulars, whitish; secondaries, chestnut, with broad tips of black; quills, bluish-green, tipped with black; under parts, blue, tinged

with green; irides, scarlet; bill, black; legs and feet,
reddish-brown.   Length, about ten inches.   The adult
female is duller in colour than the male.   The young have
the throat band indistinct, and the central tail-feathers
hardly project.   .

Roller.

THE Roller (*Coracias garrulus*, Linnæus) is a rare irregular
spring and autumn migrant to the British Isles, and a
visitant in summer to the temperate portions of Europe,
migrating into Africa in autumn.   The adult male has
the back and scapulars reddish-brown; head and nape,
greenish-blue; rump, blue; upper tail-coverts, greenish-
blue; tail, light greenish-blue, the outer rectrices with black
tips; middle feathers, darker; primaries, black, glossed with
blue; upper wing-coverts, rich blue; chin, white; remainder
of under parts, pale blue; irides, brown; bill, black; legs

and feet, warm brown. Length, about twelve inches. The
adult female resembles the male. The young are more
dingy in colour than the adults.

Iboopoe.

THE Hoopoe (*Upupa epops*, Linnæus), which is a visitor in
summer to the British Isles and to most of the temperate
portions of the Palearctic Region, passes the winter in
Central Africa and India. The adult male has the general
tint of the plumage pale reddish-buff; crest feathers, richer
in tint, and tipped with black; rump, white; tail, black,
having a broad white band in the centre; wings, black with
white bars; abdomen and under tail-coverts, white with
dark brown stripes on flanks; irides, light brown; bill,
which is slightly decurved, black; base of lower mandible,
flesh-coloured; legs, feet, and claws, dark brown. Length,

about twelve inches. The female is like the male, but the colours are not quite so bright. The young resemble the adults, but the bill is considerably shorter, and the under parts are paler in colour.

Common Kingfisher.

THE Common Kingfisher (*Alcedo ispida*, Linnæus) is a resident species in the British Islands, and is also found throughout the year in the temperate parts of the Western Palearctic Region. The adult male has the top of the head black, closely barred or spotted with bright cobalt-blue, tinged with greenish; centre of back, azure-blue; tail, deep blue; moustachial line, greenish-blue, marked with cobalt-blue, yellowish; lores and ear-coverts, chestnut; wing-coverts, dark green, with spots of cobalt; outer webs of primaries, deep blue; throat, white; under parts, yellowish-chestnut; irides, dark brown; bill, black, base, reddish; legs and feet, red. Length, from five to six inches. The adult female is duller in plumage than the male; bill, orange at base beneath.

Belted Kingfisher.

JOHN DUNCAN

THE Belted Kingfisher (*Ceryle alcyon*, Linnæus) is a very rare irregular autumn migrant to the British Islands. This bird hails from North America, and is partially migratory during winter. The general colour of the male above is slaty-blue; shaft of each feather, blackish; the head is furnished with a crest; a white spot before the eye, and a streak of the same beneath it; quills, basal half, white; secondaries and wing-coverts, tipped with white; two centre tail-feathers, bluish; remainder, brownish-black, barred with white; a broad collar of white reaches from the throat over the sides of the neck; chest band, slaty-blue; under parts, white; irides, dark brown; bill, black; legs and feet, brownish. Length, about twelve inches.

Barn Owl.

JOHN DUNCAN.

THE Barn Owl (*Aluco flammeus*, Linnæus) is a resident
in the British Isles and in most parts of Europe, but does
not occur in Norway, the north of Sweden, Russia, or on
the north-eastern shores of the Mediterranean. The adult
is buffish-orange above, spotted longitudinally with dark
brown and faint grey and white pencillings ; discs of face,

white, edged with reddish; breast, buffish, with faint dusky
spots on flanks; tail, buff, with four or five darkish
grey bars; tips of tail-feathers, white; irides, blackish;
bill, pale straw colour; claws, dark grey; legs, feathered.
Length, from twelve to thirteen inches. The female has
the same colouration as the male, but is slightly larger.

Wood Owl.

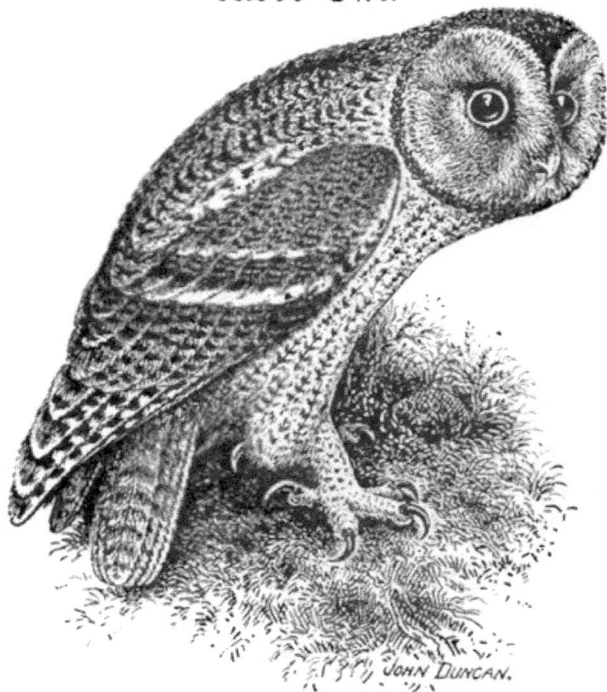

JOHN DUNCAN.

THE Wood Owl (*Strix aluco*, Gerini) is another of our
resident owls, but is not so plentiful as formerly; it is also
a resident in the Western Palearctic Region. The adult
has the upper parts reddish-brown, vermiculated and

spotted with blackish and dark brown; outer webs of wing-coverts have large subterminal white spots; facial disc and forehead, dark brown; under parts, buffish-white, streaked with dusky-brown and barred with brown; irides, black; bill, straw colour; legs and feet, feathered. Length, about fifteen inches. The female only differs in being larger. The young are usually more rufous than the adults. This owl is also subject to much variation in tint of plumage.

SHORT-EARED OWL.

THE Short-eared Owl (*Asio brachyotus*, Forster) is a regular autumn and winter visitor to the British Isles, and a few breed with us in various parts. It is also found in most parts of the Palearctic Region, as well as in the Nearctic Region. The adult has the general colour dark rich buff; wings and tail, barred transversely with dark brown; ear-tufts, short; ring round face, blackish; upper parts streaked and patched with dark brown; under parts, buffish-white, streaked longi-tudinally on breast and flanks with dusky-brown; irides, yellow; bill and claws, blackish; legs and feet, feathered. Length, from fourteen to fifteen inches. The female re-sembles the male, but is a trifle larger. (P. 159.)

Short=eared Owl.

JOHN DUNCAN.

Long=eared Owl.

JOHN DUNCAN.

THE Long-eared Owl (*Asio otus*, Linnæus) is a resident in the British Isles, also in most parts of the Western Pale-arctic Region, as well as in Central Asia. The adult has the general colour of the upper parts yellowish-buff, vermiculated and mottled with various tints of brown; ear-tufts, conspicuously long, marked in centre with black, and margined with buff; ring encircling eyes, blackish; throat, breast,

and flanks, striped with brown ; stripes on flanks, narrower ;
wings and tail, barred with brown ; irides, rich yellow ; bill,
black ; legs and feet, covered with buffish coloured downy
feathers.   Length, about fourteen inches.   The female
differs only in being larger and a little more rufous in
colour.   The young are similar to the adults.

TENGMALM'S OWL.

TENGMALM'S Owl (*Nyctala Tengmalmi*, Gmelin) is a rare
autumn migrant to the British Isles.   It is a resident
in the northern parts of the Palearctic Region, and is also
met with in Central Europe.   The adult has the upper
parts brown, more or less spotted with white ; the spots
on the top of the head are small, those on the wing-
coverts and back are larger ; tail-feathers, barred with white ;
under parts, white, washed with pale buff and greyish, and
marked with irregular bars and streaks of dark brown ;
irides, vivid yellow ; bill, straw colour ; feet, covered with
whitish feathers.   Length, about nine inches.   The female
is a little larger than the male.   (P. 162.)

Tengmalm's Owl.

JOHN DUNCAN.

Little Owl.

JOHN DUNCAN.

THE Little Owl (*Athene noctua*, Scopoli) is a rare abnormal migrant to the British Islands, and is a resident in Europe south of Scandinavia. The adult has the upper parts of the plumage greyish-brown; back and wings, spotted with white; head, striped with white; tail, barred with white and tinged with rusty; under parts, white, with broad streaks of brown; irides, yellow; bill, yellowish; the toes have a covering of hairy bristles. Length, about eight inches. The female is a trifle larger than the male. The plumage of the young is of a warmer hue than that of the adults.

Snowy Owl.

JOHN DUNCAN.

THE Snowy Owl (*Nyctea nyctea*, Linnæus) is a regular, though uncertain, autumn migrant to the British Islands.

It is a resident in the extreme north of the Palearctic and also Nearctic Regions, and in winter is met with farther south. The adult male has the whole of the plumage white, with a few spots of dark brown here and there on the upper parts; irides, rich yellow; legs and feet, clothed with long white feathers. Length, from nineteen to twenty-four inches.

HAWK OWL.

THE Hawk Owl (*Surnia funerea*, Linnæus) is an exceedingly rare wandering winter migrant to the British Islands. It inhabits the Northern Nearctic Region, and in winter moves southward to the northern portion of the United States. The adult has the general colouration of the upper plumage blackish-brown, blotched, mottled, and barred with dullish white; under parts, white, with bars of reddish-brown; tail, graduated, and marked with narrow bars of white, and broadly tipped with white; irides, pale yellow; bill, straw colour; legs and feet, covered with whitish feathers. Length, from fifteen to sixteen inches. The female is larger than the male. (P. 166.)

Hawk Owl.

JOHN DUNCAN

Scops Owl.

JOHN DUNCAN.

(See preceding page.)

THE Scops Owl (*Scops scops*, Linnæus) is a rare straggler in spring and autumn to the British Isles, and is found inhabiting the temperate parts of Europe, Western Asia, and Northern Africa. The adult is greyish above, each feather having a dark centre and vermiculated with different tints of brown; tail-feathers have bars of light and dark brown; ear-tufts, pretty well developed; under parts, grey, suffused with rich buff, streaked and finely vermiculated with brown; irides, orange-yellow; bill, black; feet, not feathered. Length, about seven inches. The female resembles the male. The young show more rufous than the adults.

Eagle Owl.

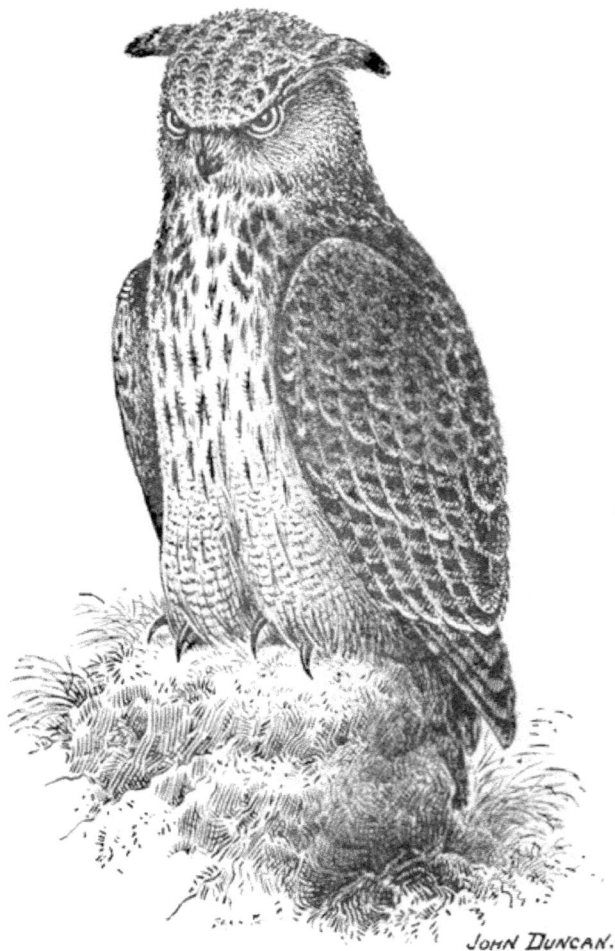

JOHN DUNCAN.

(See preceding page.)

THE Eagle Owl (*Bubo maximus*, Gerini) is a rare nomadic winter migrant to the British Islands, and is a resident in most portions of the Palearctic Region. The adult has the upper surface barred, streaked, and waved with black on a brown and yellow ground ; ear-tufts, large ; throat, white ; under parts, yellowish-buff, with longitudinal streaks of black on the chest, and fine transverse bars below ; irides, rich orange ; beak, black ; legs, feathered. Length of male, about twenty-four inches. Length of the female, about twenty-eight inches.

Griffon Vulture.

JOHN DUNCAN.

(See preceding page.)

THE Griffon Vulture (*Gyps fulvus*, Brisson) is a very rare
irregular migrant to the British Isles. It is a resident in
Southern Europe, Northern Africa, and in portions of
Central Asia ; besides being met with casually in Northern
Europe. The adult has the general colouration of the
plumage buffish-brown ; wings and tail, blackish-brown ;
some of the wing-coverts with dark centres ; under part,
reddish-brown washed with yellow ; head and neck, covered
with white down ; ruff round neck, white ; irides, deep
orange ; bill, light brown ; legs and feet, lead colour.
Length, from forty-two to forty-eight inches.

Egyptian Vulture.

JOHN DUNCAN.

THE Egyptian Vulture (*Neophron percnopterus*, Linnæus) is a very rare abnormal autumn migrant to the British Isles, and is found inhabiting European countries which border on the Mediterranean, as well as Africa and South-Western

Asia. In Northern Europe it is only of rare occurrence.
The adult is, above and below, chiefly white, tinged with
pale yellowish buff; secondaries, brownish; quills, black;
the portion of the head and neck which is bare of feathers
is of a yellowish tinge; irides, deep red; bill, dark brown;
legs and feet, flesh tint. Length, from twenty-five to twenty-
seven inches.

White Jer-Falcon.

THE White Jer-Falcon (*Hierofalco candicans*, Gmelin), also
called the Greenland Falcon, is a rare nomadic winter
migrant to the British Isles. It is supposed to be a
resident throughout most of the Circumpolar Region,
and is casually met with farther south. The adult has
the whole of the ground colour of the plumage white;
feathers on back and wing-coverts have bold spots of
blackish-brown; quills have subterminal bars of black;
tail has a few dark bars in places; under parts, spotted
here and there with blackish-brown; irides, hazel; bill,
bluish near the tip, otherwise yellowish; legs and feet,
light yellow. Length, from nineteen to twenty-one inches.
(P. 175.)

White Jer=Falcon.

Iceland Jer=Falcon.

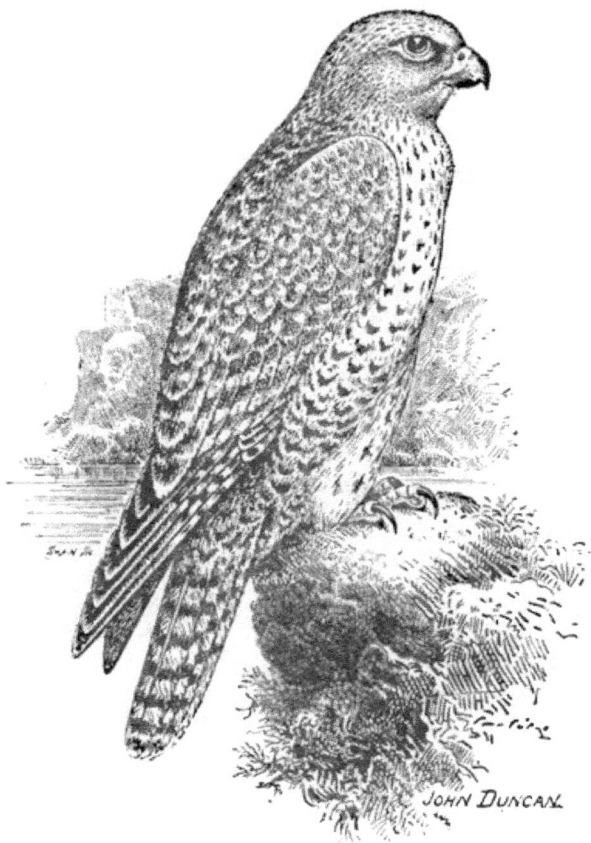

JOHN DUNCAN.

THE Iceland Jer-Falcon (*Hierofalco islandus*, Gmelin), also called the Iceland Falcon, is a rare irregular winter migrant to the British Islands, and is a resident in Iceland, and has been noticed in portions of Greenland.   The adult has the prevailing colour of the upper plumage of a creamy tint, washed with slaty-grey, and with borders and bars of a whitish colour; head, white, with thin blackish streaks; rectrices, grey, with bars of a darker tint; under parts, whitish, spotted longitudinally on breast, and barred on flanks with blackish; irides, dark brown; bill, dusky horn colour; base beneath, yellowish; legs and feet, bluish. Length, from twenty-one to twenty-two inches.

Scandinavian Jer=Falcon.

THE Scandinavian Jer-Falcon (*Hierofalco gyrfalco*, Linnæus) is a rare nomadic winter migrant to the British Isles. It inhabits Arctic Scandinavia.   This species is closely allied to the Iceland falcon, but is darker in colouration and smaller in size.

Peregrine Falcon.

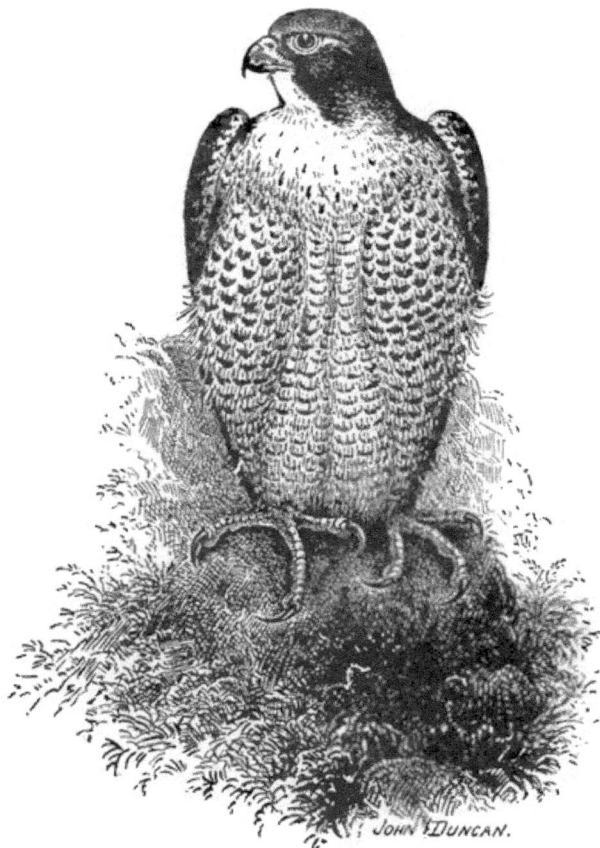

JOHN DUNCAN.

THE Peregrine Falcon (*Falco peregrinus*, Gerini) is a resident species in the British Islands; it is also met with in most parts of the Palearctic Region, but does not appear to have occurred in Iceland. The adult male has the crown,

cheeks, and moustachial region, blackish; upper parts,
bluish-grey; paler on rump, and barred with greyish-black;
primaries, blackish, with a greyish-tinge; feathers of tail,
blackish, with broad bars of bluish-grey; throat, white,
tinged with buff; breast and sides of body, crossed with
broad bars of dusky; irides, brown; bill, bluish, darker
towards the tip; cere, legs, and feet, yellow. Length, from
sixteen to nineteen inches.

Hobby.

THE Hobby (*Falco subbuteo*, Linnæus) is a regular visitor
during summer to the British Islands, and breeds throughout most parts of the Palearctic Region, spending the

winter in Africa and India. The adult male has the upper parts greyish-black, with paler margins; primaries, blackish; rump and tail-feathers, greyish, tinged with black; throat and sides of neck, white; moustachial-like streak, black; below, yellowish-white, with bold streaks of black; thighs and vent, rich yellowish-chestnut; irides, brown; bill, lead colour; cere, legs, and feet, yellow. Length, about twelve inches. The female in colouration is similar to the male, but is a little larger in size. The young have the feathers above, brown, margined with buff.

ORANGE-LEGGED HOBBY.

THE Orange-legged Hobby (*Falco vespertinus*, Linnæus) is a rare irregular spring and autumn migrant to the British Islands. It is found nesting in Central Europe and Western Asia, migrating to Africa in winter. The adult male has the principal portion of the plumage slate-grey; thighs, vent, and under tail-coverts, bright chestnut; irides, hazel-brown; bare space round eyes, orange-red; cere, orange-red; tip of bill, dark horn colour, orange at base; legs and feet, orange-red. Length, from ten to eleven inches. The adult female, which is larger than the male, has the wing-coverts, mantle, and tail ashy-grey, the two latter having blackish bars; head, nape, and under parts, rufous. The young resemble the female, but the throat is white, and under parts are streaked with brown. (P. 181.)

Orange=legged Hobby.

JOHN DUNCAN.

Merlin.

JOHN DUNCAN.

THE Merlin (*Falco æsalon*, Brisson) is a breeding species in the British Isles, but is said to be partially migratory in the winter; it also occurs in the Northern Palearctic Region. The adult male has the general colour of the upper parts slaty-blue; shafts of feathers, black; nape, rufous, streaked with dark brown; primaries, blackish, with white bars on inner webs; tail, bluish-grey, with a broad black subterminal band, and white tips; throat, white; under parts, rufous, striped with dark brown; irides, brown; bill, bluish, tip darker; cere, legs, and feet, yellow.   Length, from ten to eleven inches. The female, which is larger than the male, is deep brown above ; under parts, whitish, streaked with brown ; feathers of tail, brown, crossed with bars of deeper brown and tipped with white.

Kestrel.

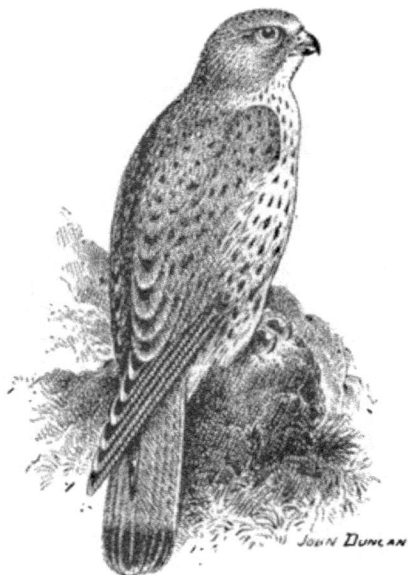

JOHN DUNCAN

THE Kestrel (*Falco tinnunculus*, Linnæus) is a resident species in the British Isles, but is partially migratory in winter, and is also found throughout the Palearctic Region (with the exception of the extreme north), also in Africa and British India. The adult male has the back light chestnut, with small spots of black; head, neck, lower back, and tail, bluish-grey, with a broad black band near the terminal end, and tipped with white; under parts, buff, spotted and streaked with black; irides, brown; bill, bluish, yellowish at the base; legs and feet, yellow. Length, from ten to thirteen inches. The female is the larger.

LESSER KESTREL.

THE Lesser Kestrel (*Falco cenchris*, Naumann) is an exceedingly rare spring and autumn migrant to the British Isles, and is found in summer on the northern shores of the Mediterranean, and to the east into Persia. It is only a straggler to Northern Europe. The adult male in colouration closely resembles *Falco tinnunculus*, but is smaller in size, and the back has no black spots, the innermost secondaries being greyish instead of rufous; irides, dark hazel; cere, yellow; bill, bluish-black at tip, yellowish at base; legs and feet, yellow; claws, generally white. Length, from about eleven to twelve inches. The adult female is similar to the adult female of the common kestrel, but is smaller in size, and the claws, as a rule, are white. (P. 185.)

Lesser Kestrel.

JOHN DUNCAN.

Golden Eagle.

THE Golden Eagle (*Aquila chrysaëtus*, Linnæus) is a comparatively scarce resident in the British Islands, and a wanderer during the non-breeding season. It is also found in most parts of the Palearctic and Nearctic Regions. The adult has the crown and nape rich tawny; back, dark brown; tail, brown, tinged with greyish, and barred obscurely with dusky-brown on basal half; tips, darker; primaries, blackish-brown; under parts, brown; thighs, dark brown; irides, bright hazel; bill, dark horn colour, base paler; cere and feet, yellow; legs feathered to the toes. Length, from thirty-one to thirty-six inches. The female is larger than the male, but similar in colour.

Spotted Eagle.

Sran Sc

John Duncan

THE Spotted Eagle (*Aquila nœvia*, Meyer) is a very rare
abnormal winter migrant to the British Isles, but is a
resident in Pomerania, moving southward through Russia,
and during migration is met with in Egypt and Northern
Africa. The adult has the general colouration dark brown;
feathers on rump, marked with light brown and whitish;
tail, dark brown, with tips of light brown; primaries, dark
brown; head and nape feathers, margined with light brown;
feathers of under tail-coverts, pale ochreous; irides, light
hazel-brown; bill, dark horn colour; tarsus, feathered; cere
and feet, yellow. Length, from twenty-four to twenty-nine
inches. The female is larger than the male, but is otherwise similar.

White-tailed Eagle.

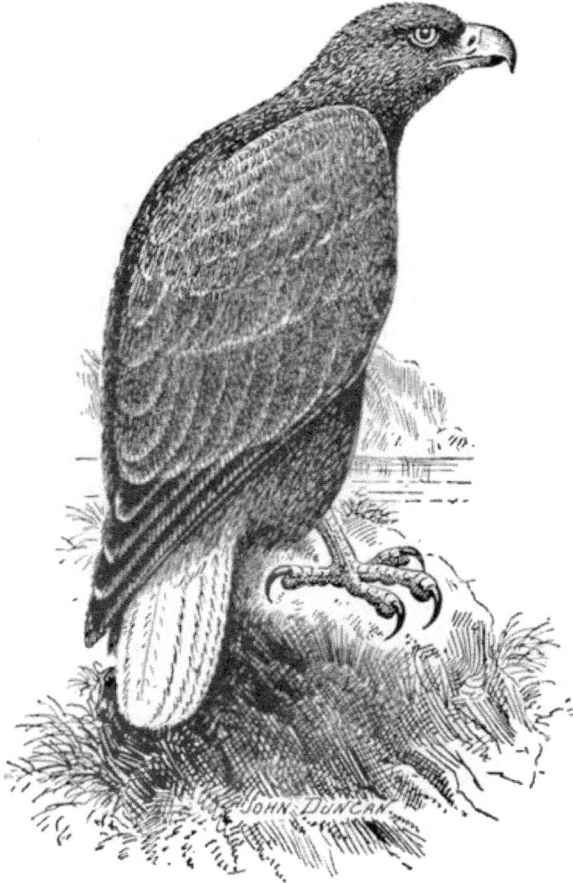

THE White-tailed Eagle (*Haliaëtus albicilla*, Brisson) is a resident in the British Isles, but, like the Golden Eagle, wanders about considerably during the non-breeding season. It is also met with in the Palearctic Region, and in South Greenland. The very old adult has the head and neck ashy-white; upper parts, brown; quills, very dark brown; tail, white; under parts, dark brown; irides, bill, cere, and feet, yellow. Length of male, about twenty-eight inches; female, about thirty-three or thirty-four inches. The females resemble males. The young have the head and tail dark brown.

Common Kite.

JOHN DUNCAN.

THE Common Kite (*Milvus regalis*, Brisson) can only be considered an accidental visitor to the British Isles, although a few pairs, probably resident birds, are occasionally found nesting in this country. It also occurs in most parts of Europe, with the exception of the extreme north and east. The adult male has the head and neck whitish, with streaks of blackish-brown; remainder of upper parts, reddish-brown; the feathers on mantle, streaked in centre with blackish; primaries, blackish; tail, forked and rufous-coloured; under parts, rufous, with blackish-brown stripes on breast; irides, yellow; bill, horn colour; cere, legs, and feet, yellow. Length, from twenty-four to twenty-six inches. The female closely resembles the male in plumage, but is a little larger in size.

Black Kite.

JOHN DUNCAN

THE Black Kite (*Milvus ater*, Gmelin) is an extremely rare abnormal spring migrant to the British Islands, and is found inhabiting Europe south of the Baltic, migrating in winter to Africa. The adult male has the upper parts of body and tail dark brown; head and neck, whitish, streaked closely with dark brown; primaries, blackish; under parts, rusty-brown, with streaks of dark brown; irides, light yellow; bill, black; cere, legs, and feet, yellow. Length, from twenty-two to twenty-three inches.

Swallow=tailed Kite.

JOHN DUNCAN

THE Swallow-tailed Kite (*Elanoides furcatus*, Linnæus) is a very rare irregular autumn migrant to the British Islands, and is found inhabiting in summer both the tropical and temperate portions of America. In winter it is migratory. The adult has the head, neck, rump, and all the under parts, white, tinged in parts with bluish-grey; rest of plumage, glossy black, with purple and blue reflections; irides, red; cere, light blue. Length, about twenty-four inches. The female is similar to the male, but is a trifle larger in size.

Honey Buzzard.

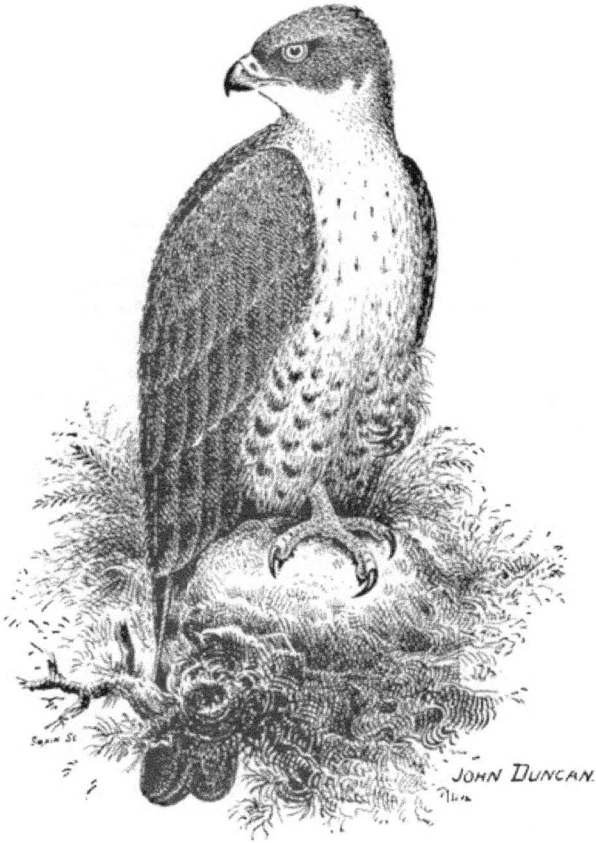

JOHN DUNCAN.

THE Honey Buzzard (*Pernis apivorus*, Linnæus) is only a summer visitor to the British Isles, and also is met with in the Palearctic and Ethiopian Regions; in the extreme north of Europe, however, it appears not to occur, and is found on the Mediterranean shores only as a migrant. The adult male has the upper parts brown; tail with three or four distinct blackish bars; head, ashy-grey; under parts, white, blotched and barred with dark brown; irides, yellow; bill, blackish; cere and feet, yellow. Length, about twenty-two inches. The female is browner on head and larger in size than male. The young is whitish on head; upper feathers, with pale edges; under parts, white with streaks of brown.

Common Buzzard.

JOHN DUNCAN

THE Common Buzzard (*Buteo vulgaris*, Leach) is generally
considered a resident species in the British Isles, but during
the non-nesting time it shifts about considerably.   It is
also found throughout Europe, but does not occur in the
extreme North or in Western Asia.   The adult has the
upper parts dark brown, margined with paler brown ; head,
pale brown, with dark brown markings ; tail-feathers, pale
brown, or greyish-white, crossed by several dark brown bars ;
quills, blackish ; chin, buffish and unspotted ; under parts,
buffish, spotted and mottled with brown ; irides, brown ;
bill, bluish-black ; cere and legs, yellow.   Length, from
twenty to twenty-three inches.

Rough=legged Buzzard.

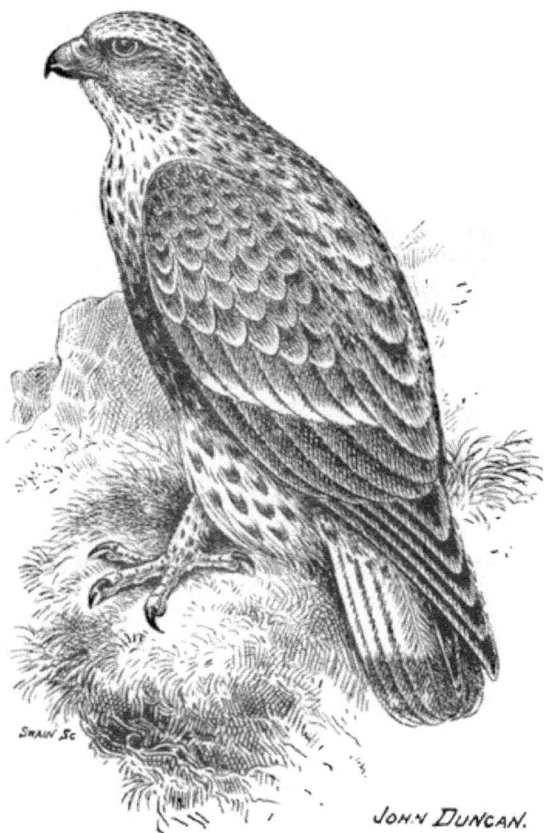

SWAIN Sc

JOHN DUNCAN.

THE Rough-legged Buzzard (*Archibuteo lagopus*, Gmelin) is
an irregular autumn visitant to the British Isles ; it is also
found inhabiting the northern portions of Europe and Asia.
The adult has the general colouration of the plumage
buffish-white, varied with brown of different shades ; back
and rump, closely marked with brown ; primaries with basal
half, white, terminal half, dark brown ; tail-feathers, white,
barred with brown, and washed with greyish-brown towards
terminal end, and tipped with whitish ; under parts, yellowish-white ; centre of abdomen, covered with a blackish-
brown patch ; irides, brown ; bill, blackish ; base, paler ;
cere and feet, yellow ; legs feathered to the toes. Length,
from nineteen to twenty-three inches. The female is the
larger.

Goshawk.

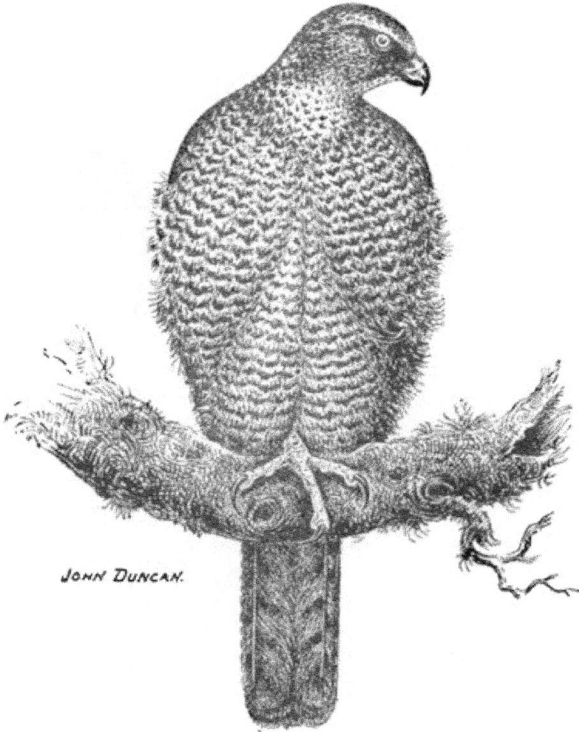

JOHN DUNCAN.

THE Goshawk (*Astur palumbarius*, Linnæus) is only an accidental visitor to the British Isles. Its range outside the British Islands extends over the whole of the Palearctic Region.   The adult has the upper parts ashy-brown; eye-stripe, lores, and nape, greyish-white ; cheeks, dark brown ; crown, dark ashy-brown ; wings and tail, greyish-brown, the latter having four bars of darker brown ; under parts,

whitish, closely spotted and barred with dark ashy-brown ;
irides, legs, and feet, yellow.   Length, from nineteen to
twenty-three inches.   The female is like the male, but larger.

Sparrow Ibawk.

JOHN DUNCAN

THE Sparrow Hawk (*Accipiter nisus*, Linnæus) is a resident
and a widely distributed species in the British Isles ; it is
also a resident throughout the whole of the Palearctic
Region.   The adult male has the upper plumage of a dark
bluish slate-colour, with a small white patch on the nape ;
tail, greyish-brown, barred with brown ; under parts, rufous,
barred with rufous of a darker shade ; irides, orange ; beak,

bluish; cere, legs, and feet, yellow. Length, about twelve inches. The female, which usually measures about three or four inches more than the male, has the upper parts of plumage brown; spot on nape, white; under parts, white, tinged with grey, and having bars of brown.

AMERICAN GOSHAWK.

THE American Goshawk (*Astur atricapillus*, Wilson) is an exceedingly rare abnormal autumn migrant to the British Islands, and inhabits the Northern Nearctic Region. During winter it is partially migratory. The adult of this species closely resembles that of *Astur palumbarius*, but is said to be darker on the head, and shows more grey on the upper parts; it is also finely freckled on the under parts, and is not barred like the common goshawk. (P. 207.)

American Goshawk.

JOHN DUNCAN.

Montagu's Harrier.

JOHN DUNCAN.

MONTAGU'S Harrier (*Circus cineraceus*, Montagu) is a migratory bird in the British Islands.   It is also found throughout most parts of Europe (with the exception of the extreme north) and Western Asia, moving southward in winter.   The adult male has the upper parts slate-grey; primaries, black; secondaries, barred with black; upper tail-coverts, white; inner web of outer tail-feathers, barred with rufous and white; chest, pale grey; remainder of under parts, white; feathers on breast, streaked narrowly with chestnut; irides, yellow; bill, black; legs, feet, and cere, yellow.   Length, from seventeen to eighteen inches. The female is usually brown on upper parts, and buffish-white below, with streaks of rufous brown.

14

Hen Harrier.

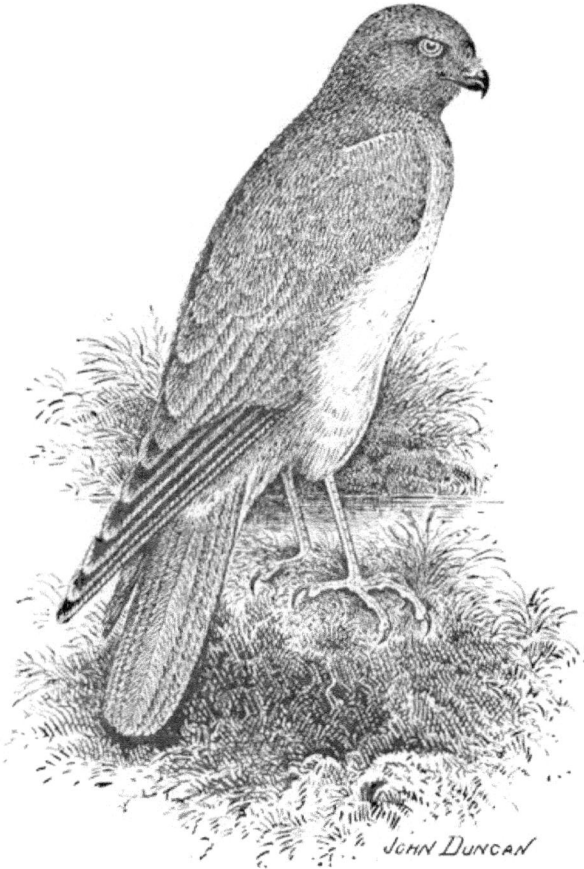

JOHN DUNCAN

THE Hen Harrier (*Circus cyaneus,* Linnæus) is for the most part a summer visitor to the British Islands. It is also met with throughout Europe, and most parts of Asia, and some portions of Northern Africa. The adult male has the upper parts light slate-grey; nape, marked with brown; rump, white; primaries, blackish; inner webs of secondaries, mostly white; throat and chest, greyish; abdomen, white; irides, cere, and legs, yellow; bill, bluish-black. Length, about eighteen inches. The female is a trifle larger than the male, and is brown above, with white streaks on nape; irides, brown; tail, brown, showing five darker-coloured bars.

Marsh Harrier.

JOHN DUNCAN.

THE Marsh Harrier (*Circus æruginosus*, Linnæus) is said to be a probable resident species in the British Islands. Throughout most parts of Europe it is a resident, but does not occur in the extreme north; it is, however, met with across Asia to Japan, and in many parts of Africa. The adult male has the head and nape white, tinged with yellowish, and with streaks of dark brown; feathers of back and scapulars, dark brown, the latter shaded obscurely with lighter brown; tail, light grey; primaries, blackish; remainder of wings, light grey; chin and throat, white, tinged with buff; rest of under parts, brown, tinged with rusty; lower part of chest has conspicuous marks of dark brown; irides, pale yellowish; bill, bluish-slate; cere, legs, and feet, yellow. Length, from nineteen to twenty-three inches.

Osprey.

JOHN DUNCAN.

THE Osprey (*Pandion haliaëtus*, Linnæus) is a summer visitor to the British Islands, and is also met with in most parts of Europe, Asia, Africa, and North America. In South America it occurs as far south as Brazil. The adult male has the head and nape white, with broad streaks of brown ; remainder of upper parts, dark brown, tinged with purplish ; beneath, white, crossed on the chest with a broad band of light brown ; irides, yellow ; bill, black ; cere, legs, and toes, bluish. Length, from twenty-one to twenty-two inches. The female is larger than the male, and has the crest more streaked and the chest band more produced.

Cormorant.

JOHN DUNCAN.

THE Cormorant (*Phalacrocorax carbo*, Linnæus) is a
resident species in the British Islands, and is also found
inhabiting most sea-coasts with the exception of those of
the Western Pacific. The adult in summer dress has the
back and wing-coverts dark greenish-brown, margined with
black; primaries and tail, black, the latter having fourteen
feathers; upper portion of head and neck, black; feathers
of crown, elongated; those on neck mixed with hair-like
feathers; throat, white; gular pouch, yellow; under parts,
rich black, a white patch on the thigh (said to be assumed
very early in spring and lost in summer); irides, green;
bill, brown; legs and feet, black. Length, from thirty-four

to thirty-six inches. The female resembles male. The young are dark brown above, with a bronze tinge; under parts, brownish-white, variegated with darker brown.

Shag.

JOHN DUNCAN.

THE Shag (*Phalacrocorax graculus*, Linnæus) is a resident in the British Isles, but shifts locally about a great deal during autumn and winter; it is also an inhabitant of Western Europe. The adult in summer has the general colour rich dark green, with reflections of bronze and purple; feathers of back, margined with black; primaries

and tail, black (the latter has twelve feathers) ; crest (which
is assumed in early spring and said to disappear by the
latter part of May), greenish-black ; under parts, deep rich
green, iridescent with purple ; irides, green ; bill, blackish ;
base of under mandible, yellow ; legs, feet, and webs, black.
Length, from twenty-six to twenty-seven inches.   The
female same as male.

Gannet.

JOHN DUNCAN

THE Gannet (*Sula bassana*, Brisson) is a resident species in
our British waters, varying in numbers according to the

time of year. It is also found inhabiting the coasts of the North Atlantic. The adult male has the head and neck white, suffused with warm buff ; the rest of the plumage, white, with the exception of the primaries, which are black ; naked skin round eyes, greyish-blue ; bill, horny-white ; irides, light straw-yellow; fore part of legs and toes, green. Length, from thirty to thirty-four inches. The female resembles male. The young, on upper parts, are dark brownish-ash, mottled with white ; under parts, dusky-ash and buff.

Mute Swan.

JOHN DUNCAN.

THE Mute Swan (*Cygnus olor*, J. E. Gmelin) is a more or less domesticated species in this country, although it is not improbable that in hard winters a few really wild examples visit the British Isles. It is an inhabitant of most portions of the Palearctic Region. The adult has the whole of the plumage white; greater part of the bill, reddish-orange; protuberance at base of upper mandible, lores, edges of mandibles, nail, and nostrils, also legs and feet, black; irides, brown. Length, about sixty inches. The female is less in size, and protuberance is not so large. The young have the plumage mostly uniform greyish-brown; bill and legs, leaden colour.

Ibooper Swan.

THE Hooper Swan (*Cygnus musicus*, Bechstein) is a common autumn and winter migrant to the British Isles, and inhabits the Northern Palearctic Region. The adult has the whole plumage white; the lores and basal portion of bill, reaching below the nostrils, yellow; remaining portion of bill, black; irides, brown; legs and feet, black. Length, about sixty inches. The female is a trifle less. The young have the upper parts ashy-brown, lighter below; bill, legs, and feet, dusky flesh colour.

Bewick's Swan.

JOHN DUNCAN.

BEWICK'S Swan (*Cygnus Bewicki*, Yarrell) is a fairly common autumn and winter migrant to the British Islands, and occurs in most parts of the Palearctic Region with the exception of Iceland. The adult has the whole plumage white; loral region and an oval patch (which does not reach the nostrils) on each side of upper mandible, yellow; rest of bill, black; irides, hazel; legs and feet, black. Length, from forty-five to forty-nine inches. The young are similar to the young of the hooper, but much less in size.

Lesser Snow Goose.

THE Lesser Snow Goose (*Chen hyperboreus*, Pallas) is a
very rare abnormal autumn migrant to the British Isles,
and inhabits the western portion of Arctic America and
North-Eastern Asia, migrating southward in winter.   The
adult has the entire plumage white with the exception of
the primary coverts, which are grey; quills, black, grey at
the base; irides, hazel;   bill, pale red;   legs and feet,

darker red. Length, from twenty-nine to thirty-two inches. The female is a little less in size than the male. The young have the upper plumage greyish; feathers on wing-coverts and back, marked in centre with darker grey; under parts, whitish; bill, black; legs and feet, pale slate colour. .

Grey=Lag Goose.

John Duncan.

THE Grey-Lag Goose (*Anser cinereus*, Meyer) is a comparatively rare resident in the British Isles, its numbers being largely increased in the winter by visitors from more northern latitudes; it also inhabits the Palearctic Region, and in winter is found in Northern Africa and India. The adult has the general colour of the head, neck, and upper parts of plumage, greyish-brown; rump and wing-coverts, bluish-grey; narrow band of white at base of upper mandible; throat and breast, grey; below, dull white; flanks, ashy-brown, margined and tipped with paler brown; irides, brown; bill, pinkish; whitish coloured nail; legs and feet, flesh tint. Length, from thirty to thirty-five inches. The female resembles male. The young are darker than mature birds.

Bean Goose.

JOHN DUNCAN

THE Bean Goose (*Anser segetum*, Gmelin) is a common autumn and winter migrant to the British Islands, and inhabits the Northern Palearctic Region. In autumn it

migrates to North-West Africa.   The adult is similar to the
grey-lag goose, but is a little less in size, the wing is longer,
and the feathers on the back have much darker brown
centres; irides are hazel; the beak is black at the base, the
middle portion orange-yellow, and the nail black; whilst
legs and feet are orange-yellow.   Length, from thirty to
thirty-four inches.

Pink-footed Goose.

THE Pink-footed Goose (*Anser brachyrhynchus*, Baillon) is
a common autumn and winter migrant to the British Isles,
and inhabits the northern portions of the Palearctic Region,
in autumn migrating southward.   The adult is distinguished
from the bean goose by its shorter wing, and the middle

15

of the bill being pink; the legs and feet are also pink.
Length, from twenty-eight to twenty-nine inches.

White=fronted Goose.

THE White-fronted Goose (*Anser albifrons*, Scopoli) is a
common, though local, autumn and winter migrant to the
British Islands. It is also found in summer in the Northern
Palearctic Region, and in autumn migrates southward to
Northern Africa and India.   The adult has the upper part
of plumage brown, tinged with ashy; feathers on back have
paler margins; base of upper mandible and forehead, white ;
primaries, black ; breast and abdomen, brownish-white, with
bold blotches and bands of black ; irides, hazel ; bill, yellow;
nail at tip, white ; legs, toes, and webs, yellow.   Length,

about twenty-seven inches. The young show no white at the base of the bill.

Lesser White=fronted Goose.

JOHN DUNCAN.

THE Lesser White-fronted Goose (*Anser albifrons minutus*, Naumann) is an exceedingly rare autumn and winter migrant to the British Islands. It inhabits Arctic Russia and Siberia, and in autumn migrates southward. The adult male closely resembles *Anser albifrons*, but is much smaller, shows more white on the forehead, and the plumage is darker ; irides, brown ; bill, whitish-pink ; nail, horn colour ; legs, toes, and webs, yellow. Length, about twenty inches.

Brent Goose.

JOHN DUNCAN.

THE Brent Goose (*Bernicla brenta*, Brisson) is a common autumn and winter migrant to the British Isles.   It inhabits the Arctic portions of the Palearctic Region, and in autumn migrates as far to the south as the Mediterranean.   The adult has the feathers on mantle blackish-brown, margined with paler ; crown, neck, chest, wings, rump, and tail-feathers, black ; on each side of the neck is a patch of black and white ; sides of rump and upper tail-coverts, white ; lower breast and belly, dark grey, margined with lighter ; vent and under tail-coverts, white ; irides, dark brown ; bill, legs, and feet, black.   Length, from twenty-two to twenty-three inches.   The young are duller in colour, and show very little white on the neck spot.

White=bellied Brent Goose.

THE White-bellied Brent Goose (*Bernicla brenta glauco-gaster*, Brehm) is an uncommon autumn and winter migrant to the British Islands. It inhabits the Arctic Regions, and migrates southward in autumn and winter. This goose is distinguished from *Bernicla brenta* by having the under parts below the breast almost white.

Bernacle Goose.

THE Bernacle Goose (*Bernicla leucopsis*, Bechstein) is a
common autumn and winter migrant to the British Isles.
It inhabits Arctic Europe, and during migration passes
southward to the Mediterranean and Egypt. The adult
has the cheeks, line over the eye, forehead, and throat,
white; lores, top of head, neck, and shoulders, black;
mantle, lavender grey, with black bars and white tips; pri-

maries and tail, blackish; breast and abdomen, whitish; sides of former and flanks have obscure bars of grey and brown; vent and tail-coverts, white; irides, hazel; bill, legs, and feet, black.   Length, about twenty-five inches.

Red-breasted Goose.

JOHN DUNCAN.

THE Red-breasted Goose (*Bernicla ruficollis*, Pallas) is an exceedingly rare abnormal autumn and winter migrant to the British Islands.   It inhabits Northern Siberia, and during migration is met with in the extreme east of Europe. The adult has the front, crown, nape, back parts of neck, patch under the eyes, and tail, black; loral patches, white; two white stripes extending from the back of the eyes to the bottom of the neck; ear-patches, neck, and upper part of the back, deep rusty-red; back and wings, dusky; lower part of breast and belly, black; sides of rump and under tail-coverts, white; flanks, barred with blackish; irides, hazel; bill, legs, and feet, black.   Length, from twenty to twenty-two inches.

Common Sheldrake.

JOHN DUNCAN

THE Common Sheldrake (*Tadorna cornuta*, S. G. Gmelin) is a resident species on the British coasts, changing quarters locally during the non-nesting period; it is also found inhabiting the Palearctic Region.   The adult male has the head and upper neck blackish-green, with glossy reflections ; lower part of neck, sides of body, lower part of back, wing-coverts, upper tail-coverts, and tail, white, the latter tipped with black ; wing spot, green ; scapulars, a portion of secondaries, and primaries, blackish ; a rich chestnut band on chest and upper back ; centre of abdomen, brownish-black ; bill and basal knob, bright red ; irides, brown ; legs and feet, red. Length, about twenty-five or twenty-six inches. The female is similar to male, but lacks knob and is a trifle less in size. The young are duller in colouration, having head and neck brown, and showing much more white than adults.

Ruddy Sheldrake.

JOHN DUNCAN.

THE Ruddy Sheldrake (*Tadorna casarca*, Linnæus) is a rare irregular autumn and winter migrant to the British Isles, and in summer inhabits the Southern Palearctic Region. The adult male has the upper back, mantle, scapulars, and the whole of the under parts, yellowish-chestnut; ring round lower neck, quills, upper tail-coverts, and tail, black, with a greenish gloss; forehead, white; head, rufous-buff; speculum, bright metallic-green; irides, dark hazel; bill, legs, and feet, black. Length, about twenty-five inches. The adult female is similar to male, but the neck ring is absent. The young in first plumage are somewhat similar to the adult females, but are lighter in colour, and the wing-coverts, scapulars, and innermost secondaries are tinged with brown.

American Wigeon.

THE American Wigeon (*Anas Americana*, Gmelin) is a very rare irregular autumn migrant to the British Islands.

It inhabits Western Arctic America, migrating southward in winter to Central America and the West Indies. The adult male has the head and neck whitish, speckled slightly with black, and a patch of metallic green on the side of the head; back and flanks, vermiculated with black on a vinous tint; axillaries, white; breast, shaded with vinous; speculum, black, glossed with green; abdomen, white; bill, slate colour; nail, black. Length, about twenty-two inches. The adult female is similar to *Anas penelope*, but the alar bar is rich black, and the axillaries white, very slightly mottled.

WIGEON.

THE Wigeon (*Anas penelope*, Linnæus) is best known as a winter visitor to the British Islands, but a few pairs nest within our limits. It also inhabits most parts of the Palearctic and Nearctic Regions, but only sparingly in the latter. The adult male has the crown and forehead ochreous-buff; cheeks and back part of neck, rich ferruginous, with minute spots of dark green; lower portion of neck behind, also scapulars and back, fine vermiculations of dark grey and white; scapulars, elongated; speculum, vivid green, tipped below with black; primaries and tail, blackish; chin, black; throat and upper part of neck, chestnut; breast, white, shading into grey; flanks, finely marked with dark grey; irides, hazel; bill, slate-grey; tip, black; legs and feet, deep greyish-blue. Length, about twenty or twenty-one inches. (P. 237.)

Wigeon.

JOHN DUNCAN

Common Teal.

JOHN DUNCAN

THE Common Teal (*Anas crecca*, Linnæus) is a locally
distributed species throughout the British Islands, and also
inhabits the Palearctic Region, but is rare in the Nearctic
Region.   The adult male has the crown, nape, and cheeks
bright chestnut; enclosing the eye, and reaching the nape,
is a large patch of bright green with glossy purple reflections; bordered by black and white; sides of lower neck,
back, scapulars, and flanks, finely vermiculated with black
and white; wing-coverts, brown, shaded with greyish;
speculum, black, green, and purple, tipped with pale buff;
rump and tail-coverts, blackish; feathers of tail, brown,
washed with ash; chin, black; breast, chestnut, spotted
with round black spots; middle of abdomen, white; under
tail-coverts, black in the centre with rich buff on each side;
irides, hazel; bill, black; legs and feet, brownish.   Length,
about fourteen inches.

American Teal.

THE American Teal (*Anas Carolinensis*, Gmelin) is a very rare abnormal autumn migrant to the British Islands, and is found inhabiting North America, migrating south in winter. The adult has the scapulars pale slate-grey, and a broad white crescent on each side of the breast.

Blue-winged Teal.

JOHN DUNCAN

THE Blue-winged Teal (*Anas discors*, Linnæus) is an exceedingly rare straggling autumn migrant to the British Islands, whose claims to be included in the British list are very doubtful. It is found inhabiting the Central and Southern Nearctic Region. The adult male of this teal can readily be distinguished from its allies by a white crescent between the eye and the bill, and by the conspicuous blue wing-coverts.

Gadwall.

THE Gadwall (*Anas strepera*, Linnæus) is a resident species in some parts of the British Islands, nesting in some districts in fair numbers.   It occurs in both the Palearctic and Nearctic Regions.   The adult male has the head and upper part of neck pale brown, mottled with darker brown; back and lesser wing-coverts, dark brown, edged with pale grey; upper and under tail-coverts, black; median coverts, chestnut; greater coverts, blackish; primaries, brown; speculum, white; chest and flanks, dusky, margined with paler; centre of abdomen, whitish; irides, brown; bill, black; legs and feet, dull orange-red.   Length, from nineteen to twenty inches.   The female has the upper parts of plumage and breast margined with pale brown; speculum, white. The young are brown of various shades; speculum, white.

Pintail Duck.

THE Pintail Duck (*Anas acuta*, Linnæus) is a fairly common visitant to the British Isles in winter. It is also met with in the Northern Palearctic and Nearctic Regions, and during winter is found in the Oriental Region. The adult male has

16

the back and scapulars (the latter being elongated) finely
vermiculated with brown and grey; speculum, vivid green,
margined with black and white; tail, blackish-grey; central
pair of rectrices, black, and elongated; under tail-coverts,
black; head and throat, dark reddish-brown; back of neck,
darker and colours glossed; a white stripe runs down each
side of neck; under parts, white; irides, brown; bill,
leaden-blue, black in upper centre; legs and feet, black.
Length, about twenty-four inches.   The female has no
long tail-feathers.

GARGANEY.

THE Garganey (*Anas circia*, Linnæus) is a summer migrant
to the British Isles; it inhabits most parts of the Palearctic Region with the exception of the extreme north.   In
winter it occurs as far southwards as the Malay Archipelago.
The adult male has the back brown, glossed with green,
and paler margins; scapulars, long and elongated, and
conspicuously marked in centre with a white stripe; wing-
coverts, bluish, tinged with grey; speculum, glossy green,
with white margins on either side; quills and tail, brown;
crown and nape, deep rich umber-brown, with a border of
white; chin, black; cheeks and throat, rufous-brown;
breast, light brown, with crescentic bands of dark brown;
abdomen, white; with fine blackish vermiculations towards
vent, and having two black crescents on each lower flank;
irides, hazel; bill, blackish; legs and feet, greyish-brown.
Length, about fifteen inches.   (P. 243.)

Garganey.

Shoveller.

JOHN DUNCAN

THE Shoveller (*Anas clypeata*, Linnæus) is principally a winter visitor to the British Isles, but a few are always found in the nesting season breeding with us. It also inhabits both the Palearctic and Nearctic Regions. The adult male has the centre of back brown, margined with paler; rump and tail-coverts, rich black; tail, black; head and fore part of neck, rich metallic-green; majority of feathers on lower neck and scapulars, white; elongated scapulars, black, blue, and white; speculum, bright green; greater wing-coverts, tipped with white; primaries, black; breast and abdomen, bright rich chestnut; vent, white; under tail-coverts, black; irides, yellow; bill, bluish-slate, and dilated at the end; legs and feet, orange. Length, about twenty-one inches. The female has the upper parts deep brown, margined and barred with reddish-white; under parts, light reddish-brown.

Mallard.

THE Mallard (*Anas boschas*, Linnæus) is a resident species in the British Isles, and also inhabits most portions of the Palearctic and Nearctic Regions. The adult male has the head and neck glossy green; followed by a collar of pure white; breast, deep chestnut-red; mantle, chestnut-brown, feathers margined with paler; scapulars, greyish-white, ver-miculated with brown; rump and centre tail-feathers, which are curled up, black; lesser wing-coverts, greyish-brown; greater coverts, barred with white and tipped with black; speculum, brilliant violet, shading into black and tipped with white; primaries, greyish-brown; abdomen and flanks, white tinged with grey, and finely vermiculated with dusky; irides, dark brown; bill, yellowish; legs and feet, orange. Length, about twenty-four inches.   The female has the general colouration brown and buff of various shades.

Pochard.

THE Pochard (*Fuligula ferina*, Linnæus) is a fairly common winter visitor to the British Isles, and it also inhabits the Palearctic Region, but does not occur in the extreme North and East.   The adult male in summer dress has the head and upper neck rich chestnut; breast and upper portion of back, brownish-black; mantle and breast, white, tinged

with grey, with fine vermiculations of dark grey; upper tail-
coverts, black; speculum, grey; under parts, greyish-white,
vermiculated with grey; under tail-coverts, black; irides,
red; bill, black, showing a broad band of blue across the
centre; legs and toes, bluish-grey; webs, black. Length,
about eighteen inches. The female has the chin whitish;
head, neck, and breast, brownish. The remainder of
plumage shows more brown than in the male.

Red=crested Pochard.

THE Red-crested Pochard (*Fuligula rufina*, Pallas) is a
rare irregular winter migrant to the British Isles, and also

inhabits the South-Western Palearctic Region. The adult male has the head and fore-neck rufous, the former being furnished with a soft bushy crest; lower neck, breast, and belly, rich warm black; secondaries, white; some of the quills and speculum, white; back and tail, brown; shoulders and flanks, white; irides, red; bill, rich crimson; nail, brown; legs and feet, rich orange. Length, about twenty-one inches. The adult female has the general colour of the upper parts greyish-brown; head, tinged with yellowish-brown; white shoulder patches absent.

WHITE-EYED POCHARD.

THE White-eyed Pochard (*Fuligula nyroca*, Guldenstadt) is a rare abnormal spring, autumn, and winter migrant to the British Isles, and is also found in the Southern Palearctic Region and North-Western Oriental Region. The adult male has the back, wings, and tail deep brown; a white speculum with a black border in front; head, neck, and breast, rich chestnut; a dusky ring encircles the lower neck; a small white spot on chin; abdomen, whitish; flanks, dull chestnut; under tail-coverts, white; irides, white; bill, blackish; nail, black; legs and feet, slate colour. Length, about sixteen inches. The adult female closely resembles the male, with the exception that the colouration is duller and the belly marked with brown. The young of the year are similar to the adult females, but show less white and chestnut. (P. 249.)

White=eyed Pochard.

Tufted Duck.

THE Tufted Duck (*Fuligula cristata*, Leach) is a winter visitor to the British low-lying coasts, and a fair number are known to nest with us; it also occurs in the more temperate portion of the Palearctic Region. The adult male has the head, elongated crest, and upper portion of neck, black, glossed with purple; lower part of neck, black; mantle and scapulars, brownish-black, with fine vermiculations of brown; speculum, white; abdomen, white, tinged with grey towards vent; under tail-coverts, black; irides, light yellow; bill, bluish-grey, with black nail; legs, bluish-grey; webs, dusky. Length, about fifteen inches. The female has the parts which are black in the male, brownish; abdomen and sides, white, with a greyish tinge; on forehead, some white mottlings.

Scaup.

THE Scaup (*Fuligula marila*, Linnæus) is a common autumn and winter migrant to the British Islands, and inhabits both the Northern Nearctic and Palearctic Regions. The adult male has the middle of the back and scapulars silvery white, with black vermiculations; speculum and under parts, white; head, neck, and breast, black, glossed with purple ; most feathers of wings, greyish-brown ; rump, upper and under tail-coverts, black, with glossy reflections ; irides, yellow ; bill, pale slate-blue ; nail, black ; legs and feet, slate-grey. Length, from seventeen to nineteen inches. The adult female has the general colour of upper plumage and breast brown, with a few vermiculations of whitish ; flanks, faintly barred with brown ; forehead and chin, whitish. The young resemble the adult females, but the feathers round the base of the bill show much less whitish.

GOLDEN EYE.

THE Golden Eye (*Clangula glaucion*, Linnæus) is a common autumn and winter migrant to the British Islands, and inhabits the Northern Nearctic and Palearctic Regions. The adult male has the head and upper part of neck black, with green and purple reflections ; a white spot under the lores ; upper plumage varied with black and white ; under parts, white ; thighs, brown ; irides, rich yellow ; bill, black ; legs and toes, orange ; webs, blackish. Length, from seventeen to nineteen inches. The adult female is less in size, and lacks the white on the face. The young in first plumage are similar to the adult female. (P. 253.)

Golden Eye.

JOHN DUNCAN.

Harlequin Duck.

THE Harlequin Duck (*Fuligula histrionica*, Linnæus) is an exceedingly rare nomadic autumn migrant to the British Isles, and inhabits the Eastern Palearctic and Nearctic Regions.   The adult male of this duck cannot be mistaken for any other British species, and a detailed description is hardly needed.   The general colouration of the plumage is of a leaden tint, varied with markings of white, black, chestnut, and brown; the wing speculum is metallic purple; irides, hazel; bill, dark slate, lighter on the nail; legs and feet, brown, with darker webs.   Length, about seventeen inches. The adult female is mostly uniform brown, with a faint whitish patch between the base of the bill and eye; under parts, mottled with white.

Long=tailed Duck.

THE Long-tailed Duck (*Fuligula glacialis*, Linnæus), a tolerably common autumn and winter migrant to the British Islands, is a circumpolar species, migrating in autumn southward through both Eastern and Western

Hemispheres. The adult male in summer has the head, neck, upper part of breast, and upper back, white; eyes, surrounded with a greyish tinge; a large patch of dark brown on each side of lower scapulars, elongated and white; back, rump, wings, tail and upper tail-coverts, deep brown; two central tail-feathers which are narrow and much elongated, brown; outer rectrices, white; flanks, washed with grey; breast, dark brown; below, white; irides, reddish-brown; bill at basal half and nail, black; remainder, orange-red; legs and feet, deep slate-grey. Length, without central tail-feathers, about twenty-two inches. The adult female lacks the elongated rectrices.

COMMON SCOTER.

THE Common Scoter (*Fuligula nigra*, Linnæus) is an autumn and winter visitor to the British Islands, breeding very locally in Scotland; it is also found inhabiting the whole of the Northern Palearctic Region. The adult male has the whole of the plumage rich glossy black; irides, hazel; bill, black, with central ridge of upper mandible deep yellow; legs and feet, blackish, webs of a darker hue. Length, from eighteen to twenty inches. The adult female has the prevailing colouration of the plumage of a brownish tinge; knob on bill hardly perceptible. The young are similar to adult female, but under parts are mottled with whitish. (P. 257.)

Common Scoter.

Velvet Scoter.

JOHN DUNCAN

THE Velvet Scoter (*Fuligula fusca*, Linnæus) is a fairly common autumn and winter migrant to the British Islands. It inhabits the Northern Palearctic Region, and is met with on migration as far south as the Mediterranean. The adult male has the whole of the plumage black (glossed above, and duller beneath), with the exception of a spot behind the eye and speculum, which is pure white ; irides, brownish ;

bill, black on the basal tubercule and nostrils; margin of
upper mandible, also black; remainder, orange; legs and
feet, orange; webs, blackish-brown.   Length, about twenty-
two inches.   The adult female is browner on the upper
parts; underneath, greyish, streaked and spotted with
brown; speculum, whitish; a whitish spot on lores, and a
patch of whitish on auriculars.

Surf Scoter.

JOHN DUNCAN

THE Surf Scoter (*Fuligula perspicillata*, Linnæus) is a rare
straggler in autumn and winter to the British Isles, and
inhabits the Northern Nearctic Region.   The adult male in
summer has the plumage velvety black, with a tinge of
brown on the throat; a broad white band between the

eyes, and a triangular patch of the same colour on the nape; irides, straw colour; bill, orange-red on the upper mandible; nail, yellowish-grey; protuberance on each side at the base, black; and in front a silvery-greyish patch extending to the nostrils; lower mandible, pinkish; legs and feet, orange-red; webs, dusky. Length, about twenty-one inches. The adult female is of a uniform sooty-brown colour, lightest about the neck, and the protuberances are scarcely to be seen; whilst the colour is dusky.

Buffel-headed Duck.

The Buffel-headed Duck (*Clangula albeola*, Linnæus) is a very irregular winter migrant to the British Islands, and inhabits the Northern Nearctic Region. The adult is con-siderably less in size than *Clangula glaucion*, to which it bears a close resemblance. It can readily be distinguished

by the large white patch on the side of the head behind
the eye; the common golden-eye has the white in front of
the eye.   The adult female has also a white patch behind
the eye.

Common Eider.

JOHN DUNCAN.

THE Common Eider (*Somateria mollissima*, Linnæus) is a
resident species in the British Isles, but is only known to
breed in England in one locality—viz., the Farne Islands,
Northumberland.   It is also found inhabiting the northern
portions of the Western Palearctic Region.   The adult

male has the forehead and crown black, with a line of white
on the hind crown ; nape, emerald green, with a divisional
line of white between a green patch on each side of the
neck ; throat, white ; upper breast, rich buff; under parts,
black, showing a white spot on each side of vent ; lower
back, rump, upper tail-coverts, primary-coverts, greater
wing-coverts, and secondaries, black ; quills and tail, dark
brown ; irides, hazel-brown ; bill, olive-green ; legs and feet,
olive-green.   Length, from twenty-four to twenty-six inches.
The female is principally brown and black.

STELLER'S EIDER.

STELLER'S Eider (*Somateria stelleri*, Pallas) is an exceedingly rare nomadic autumn and winter migrant to the British
Isles.   It inhabits the North-Eastern Palearctic Region, and
probably the extreme North-Western Palearctic Region.
The adult of this eider is easily distinguished by a black
ring (which is glossed with purple and green) round the
neck ; the elongated secondaries, which are white on the
inner and bright blue on the outer webs ; the wing-speculum,
which is bluish-purple ; and a black spot on each side of
the breast.   Irides, hazel ; bill, dark slate ; nail, lighter ;
legs and feet, brownish-grey.   Length, from eighteen to
nineteen inches.   The adult female has the upper parts
brown ; under plumage, blackish on abdomen ; speculum,
dull bluish-purple.   (P. 263.)

Steller's Eider.

JOHN DUNCAN.

King Eider.

THE King Eider (*Somateria spectabilis*, Linnæus) is a rare accidental straggling autumn and winter migrant to the British Isles, and is found inhabiting the Circumpolar Region, during winter moving southward.　The adult male has the crown and nape light bluish-grey; cheeks, pale emerald green; line above the eye and breast, tinged with buff; lower portion of neck, upper back, and wing-coverts, white ; scapulars and innermost secondaries, black ; primaries, brown ; rump, tail-coverts, and lower parts (except a white patch on flanks), black ; irides, yellow ; bill and naked elevated basal tubercule, rich orange, the latter bordered with black ; upper throat has a V-shaped mark of black ; legs and feet, orange. Length, from twenty-two to twenty-four inches.

Hooded Merganser.

THE Hooded Merganser (*Mergus cucullatus*, Linnæus) is a rare nomadic autumn and winter migrant to the British Islands, and is also found inhabiting the Northern Nearctic

Region.   The adult male of this merganser can readily be
distinguished by the semicircular black crest and the broad
bar or patch of white behind the eye.   Length, about nineteen inches.   The adult female has the crest reddish-brown,
and is rather smaller in size than the male.

GOOSANDER.

THE Goosander (*Mergus merganser*, Linnæus) is a winter
visitor to the British Isles, and has nested in some parts
of the Highlands.   It also occurs in the Palearctic and
Nearctic Regions.   The adult male has the head and upper
part of neck black, glossed with green and purple ; feathers
on crown and nape, long and silky ; lower part of neck and
remainder of under parts, white ; breast and abdomen,
suffused with rich buff-orange ; thighs, slightly vermiculated
with grey ; mantle, black ; wing-coverts and outer secondaries, white ; quills, black ; rump and tail, ashy-black ;
irides, red ; bill, vermilion ; nail, black ; legs and feet, rich
orange.   Length, about twenty-six inches.   The female is
chestnut on head and upper neck ; chin and upper throat,
whitish ; feathers on crown and nape, elongated ; upper
parts, slate-grey ; greater coverts, conspicuously tipped with
white ; under parts, whitish.   (P. 267.)

Goosander.

DUNCAN

Red=breasted Merganser.

JOHN DUNCAN

THE Red-breasted Merganser (*Mergus serrator*, Linnæus) is a winter visitant to England, but nests in both Scotland and Ireland, and is also found in the Northern Palearctic and Nearctic Regions.   The adult male in summer has the head, crest, and upper portion of neck greenish-black, with purple reflections ; a black line runs down the back of the neck ; near the point of the wing is a tuft of white feathers with broad black edgings; mantle, portion of shoulders, inner scapulars, basal half of greater wing-coverts, secondaries, and primaries, rich black ; speculum, white ; long tertials bordered with black ; white collar round neck ; upper part of breast, pale chestnut-brown, streaked with black ; remainder of under parts, white ; flanks, rump, and tail-coverts have grey vermiculations ; irides, red ; bill, red ; nail, black ; legs and feet, reddish-orange.   Length, about twenty-two inches.   The adult female is much less than the male, and is reddish-brown on the head and neck ; a black bar runs across the alar speculum.

Smew.

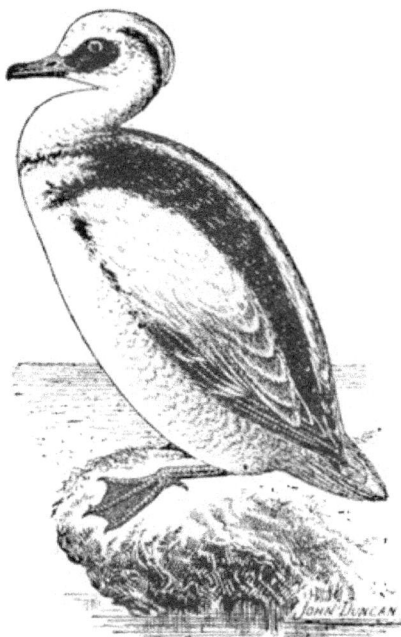

THE Smew (*Mergus albellus*, Linnæus) is a rare straggling autumn and winter migrant to the British Isles, and inhabits the Northern Palearctic Region. The adult male has the forehead, crown, and crest (which is elongated) white; loral region, black; patch on nape, black; back, two crescentic black lines on sides of breast; secondaries and middle wing-coverts, rich black, the latter fringed with white; primaries, blackish; scapulars, white and greyish, with black margins on outer webs; upper tail-coverts and tail, slate-grey; flanks and sides, vermiculated with black; plumage, otherwise white; irides, reddish; bill, slate colour, lighter at tip; legs and feet, grey; webs, darker. Length, from sixteen to seventeen inches.

Flamingo.

JOHN DUNCAN.

THE Flamingo (*Phœnicopterus roseus*, Pallas) is an exceptionally rare abnormal migrant to the British Islands, and is found inhabiting the South-western Palearctic Region, as well as the Northern Ethiopian Region. The adult has the wing-coverts scarlet ; quills, black ; remainder of plumage, white, washed with rose colour ; irides, yellow ; base of bill, rosy, tip, black ; legs and feet, light pink. The length is said to vary from forty-four to seventy inches.

GLOSSY IBIS.

THE Glossy Ibis (*Plegadis falcinellus*, Linnæus) is a rare abnormal spring and autumn migrant to the British Islands, breeding in Southern Europe, and distributed over most of the temperate and tropical regions of both hemispheres. The adult has the general colouration, above, glossy greenish-black, with a metallic lustre ; head, neck, and under parts, chestnut ; irides, brown ; naked skin, from base of bill and round eyes, green ; bill and legs, greenish-black. Length, about twenty-two inches. The male and female are alike in plumage, but the latter is less in size. The young are duller in colour, and streaked in parts with greyish-white. (P. 274.)

Glossy Ibis.

Spoonbill.

JOHN DUNCAN.

(See preceding page.)

THE Spoonbill (*Platalea leucorodia*, Linnæus) is an occasional straggler to the British Islands during migration,
and is found inhabiting the Palearctic and Western Oriental
Region, migrating southward in winter.   The adult male
has the whole of the plumage white, with the exception of
the crest and feathers on lower neck, which are slightly
suffused with pale yellow; irides, red; bill, tipped with
yellow; remainder, black with yellow bars; naked spot on
throat, yellow; legs and feet, black.   Length, about thirty-
two inches.   The adult female is a trifle less, and the crest
is not so much developed.

WHITE STORK.

THE White Stork (*Ciconia alba*, Brisson) is a rare irregular
spring and autumn migrant to the British Isles, and inhabits
the Western Palearctic Region.   The adult has the whole
of the plumage white, with the exception of the secondaries
and quills, which are black; irides, brown; naked space
round eye, black; bill, legs, and feet, red.   Length, from
forty to forty-four inches.   (P. 277.)

White Stork.

JOHN DUNCAN.

Black Stork.

JOHN DUNCAN

THE Black Stork (*Ciconia nigra*, Linnæus) is a very rare
spring and autumn migrant to the British Isles, and is
found inhabiting the Southern Palearctic Region.   The
adult has the under parts from the lower breast, white ;
remainder of plumage rich black, more or less suffused with
purple, blue, coppery, and green reflections ; irides, brown ;
bare space round eye, scarlet ; bill, legs, and feet, scarlet.
Length, from forty to forty-two inches.

Bittern.

JOHN DUNCAN.

THE Bittern (*Botaurus stellaris*, Linnæus) can only be called a casual visitor to the British Isles in spring. At one time it nested with us, but is said no longer to do so. It is found also in most parts of the Eastern Hemisphere, with the exception of the extreme north. The adult has the upper parts more or less buff, with verniculations of brown and bars of dark brown; forehead, crown, and nape, brown; side of chin, streaked with blackish; rest of plumage, more or less vermiculated, streaked, and marked with chestnut, buff, and brown; irides, brown; bill, greenish-yellow; legs and feet, greenish. Length, from twenty-five to twenty-eight inches. The female is similar to the male. The young resemble adults.

Little Bittern.

THE Little Bittern (*Ardetta minuta*, Linnæus) is a spring
and autumn migrant to the British Islands, and is found
inhabiting Central Europe and Western Asia, ranging to
the south in winter as far as Central Africa. The adult
male has the crown, nape, back, shoulders, and tail-feathers
glossy greenish-black; quills, black; sides of head, neck,
and wing-coverts, rusty-buff, the latter paler; throat and
under parts, buff, streaked with brownish on some of the
feathers at side of chest and flanks; chin and vent, white;
irides and bill, yellow; legs and feet, yellowish-olive.
Length, from twelve to thirteen inches. The adult female
is a little less than the male.

American Bittern.

JOHN DUNCAN.

THE American Bittern (*Botaurus lentiginosus*, Montagu) is a rare abnormal spring and autumn migrant to the British Islands, and inhabits the Nearctic Region, but not the extreme north. The adult is similar in general colouration of plumage to the common bittern. It is, however, smaller in size, the vermiculations are much finer, and the primaries, which are uniform slate-brown, are not barred. The bill, legs, and feet are also more slender than those of *Botaurus stellaris*. Length, about twenty-seven inches. The female resembles the male, but is less in size. The young show more rufous in the colouration.

COMMON HERON.

THE Common Heron (*Ardea cinerea*, Linnæus) is a resident in the British Islands, and is found inhabiting most parts of Europe, Asia, and Africa, with the exception of the extreme north of the two former continents. The adult male has the forehead, crown, sides of head, throat, edge of wings, breast, abdomen, and lower tail-coverts, white ; upper parts, mostly ashy-grey, with the elongated feathers lighter grey ; primaries, black ; sides of breast, boldly marked with black ; neck, white, streaked in front with blackish ; occipital plumes, much elongated and black ; irides, yellow ; bill, yellow ; legs and feet, yellowish-green. Length, about thirty-six inches. The adult female closely resembles male, but the plumage is duller and crest shorter. (P. 284.)

Common Heron.

JOHN DUNCA.

Night Heron.

THE Night Heron (*Nycticorax griscus*, Linnæus) is a rare spring and autumn migrant to the British Isles, and inhabits the Southern Palearctic Region, as well as Africa. The adult has the crown, nape, upper back, and scapulars deep brown, with a metallic-green gloss; wings and tail, pale slate; forehead, eye-stripe, occipital plumes, and remainder of plumage, white; irides, deep red; bill, blackish; legs and feet, yellowish-brown. Length, from twenty-two to twenty-three inches. The female has the occipital plumes less developed, and the plumage is duller in colouration than the male. The young bird has no crest.

Buff=backed Heron.

John Duncan.

THE Buff-backed Heron (*Ardea bubulcus*, Audouin) is an
exceedingly rare abnormal spring and autumn migrant to
the British Isles. It inhabits Southern Europe and Africa,
but is met with in the south-western parts of Asia. The
adult male in summer has the plumage of the head, neck,
breast, and the elongated filamentous feathers of plumes,
rusty-buff; remainder of plumage, white; irides, yellowish-
pink; bill, reddish at base, yellow at tip; legs, yellowish;
lores, yellow. Length, from eighteen to twenty inches.
The female resembles the male, but is rather less, and the
plumes are smaller.

Squacco Heron.

JOHN DUNCAN.

THE Squacco Heron (*Ardea comata*, Pallas) is a rare ab-normal spring and summer migrant to the British Islands. It is found inhabiting Southern Europe and Africa, and in winter migrates from Europe. The adult has the feathers of cheeks, back, neck, and upper breast, buff, tinged with light reddish on the back; feathers on the crown, which are much lengthened, white, bordered with black; dorsal plumes, much elongated; remainder of plumage, white; irides, light yellow; bill, blue, black at tip; bare part round eye, green; legs and feet, yellowish flesh colour. Length, from eighteen to nineteen inches. The adult female is similar to the adult male, but the plumes are less developed.

Purple Heron.

JOHN DUNCAN.

THE Purple Heron (*Ardea purpurea*, Linnæus) is a rare irregular spring and autumn migrant to the British Isles. It inhabits most of the temperate and tropical regions of the Eastern Hemisphere; in the northern parts it is migratory. The adult has the crown and elongated feathers of occiput purplish-black; back and wing-coverts, deep slate-grey; dorsal plumes, rusty-red; chin and throat, white; sides of head and neck, ferruginous, with streaks of blackish; neck, yellowish-chestnut, with a black line down each side; elongated feathers on chest, black, grey, and reddish; under parts, rich chestnut; irides, yellow; bill, yellow, brownish above; legs and feet, olive. Length, from twenty-nine to thirty-six inches.

LITTLE EGRET.

THE Little Egret (*Ardea garzetta*, Linnæus) is an exceedingly rare irregular spring and autumn migrant to the British Isles, and inhabits the Southern Palearctic Region and Ethiopian and Oriental Regions. The adult in summer plumage has a crest on the occiput consisting of two elongated feathers; lower neck, furnished with lengthened feathers; back plumes, elongated; entire plumage, white; irides, light yellow; naked part round eye, lavender; bill and legs, blackish; lower part of feet, yellowish. Length, about twenty-four inches. The plumes of the adult female are said to be shorter than those of the adult male, but otherwise the plumage is similar. The young have no lengthened plumes. (P. 290.)

Little Egret.

JOHN DUNCAN.

Great White Egret.

THE Great White Egret (*Ardea alba*, Linnæus) is a very rare abnormal spring and autumn migrant to the British Isles, and is found inhabiting the Southern Palearctic Region.   The adult in summer has a short occipital crest ; dorsal plumes, which are elongated, hang over the tail ; feathers at bottom of neck, prolonged and lancelote ; plumage, entirely white ; irides, yellow ; bare part round eye, greenish-yellow ; bill, black ; legs and feet, reddish-brown.   Length, from forty to forty-two inches.   The plumes, which are shorter in the adult female, are absent in the young birds ; bill, yellow.

Common Crane.

JOHN DUNCAN.

THE Common Crane (*Grus communis*, Bechstein), which is
a rare straggling spring and autumn migrant to the British
Isles, is found inhabiting the Palearctic Region. The adult
has the general colour grey; top of the head, which is bare
of feathers, red; forehead and lores, black; nape, chin, and
front of neck, greyish-black; primaries, black; innermost
secondaries, elongated and forming graceful plumes; irides,
crimson; bill, olivish; legs and feet, blackish. Length,
from forty-five to forty-six inches.

DEMOISELLE CRANE.

THE Demoiselle Crane (*Grus virgo*, Linnæus) is an exceedingly rare straggling spring migrant to the British
Isles, and inhabits the Southern Palearctic Region. The
adult male has the general colouration ashy-grey; the sides
of the head, neck, and a prolonged plume depending from
the breast, blackish; behind each eye is a long tuft of white
feathers passing backwards to the occiput; secondaries,
elongated into slender-pointed plumes, which fall over quills
and tail; irides, crimson; bill, greenish at base, light brown
at tip; legs and feet, blackish. Length, about thirty-six
inches. The female is duller in colour than the male, and
the elongated feathers are less developed. (P. 294.)

Demoiselle Crane.

JOHN DUNCAN.

Great Bustard.

Smith Sc   JOHN DUNCAN.

THE Great Bustard (*Otis tarda*, Linnæus), which formerly
bred in, but is now only an irregular nomadic spring,
autumn, and winter migrant to, the British Isles, inhabits
the Southern Palearctic Region. The adult male has the
head pale grey; on each side of the base of the bill is a
long tuft of whitish feathers; throat and upper neck,
white; upper plumage, mostly yellowish-rusty or buffish-
red, barred and otherwise variegated with brown and
black; chest, yellowish-rusty, with markings of chestnut;
underneath, white; irides, dark hazel; bill, grey; tip,
darker; legs and feet, brown. Length, about forty-five
inches. The adult female is much smaller, and the moustachial tuft and chestnut markings on chest are absent.

Little Bustard.

THE Little Bustard (*Otis tetrax*, Linnæus), which is a rare straggling spring, autumn, and winter migrant, inhabits the South-western Palearctic Region. The adult male of this bustard bears a close resemblance in the colouration of the plumage to that of *Otis tarda*, but the chest is crossed by two black bands, and the moustachial plumes (which are conspicuous in the great bustard) are absent; irides, bright yellow; bill, brownish; tip, darker; legs and feet, dusky-yellow. Length, from sixteen to seventeen inches.

Macqueen's Bustard.

JOHN DUNCAN.

(See preceding page.)

MACQUEEN's Bustard (*Otis Macqueeni*, Gray) is an exceedingly rare abnormal autumn migrant to the British Isles,
and inhabits the South Central Palearctic Region.   The
adult male has the upper plumage light sandy-buff and rich
cream or reddish-fawn colour, with darker markings here
and there, and fine vermiculations of dark brown ; the
upper crest feathers are white tipped with black ; undermost
crest feathers, white ; ruff on side of neck, mostly black ;
tail, barred irregularly with three bands of bluish-grey ; sides
of head, pale fawn, with fine vermiculations of brown ;
feathers of crop, grey and elongated ; under parts, white ;
under tail-coverts, banded and vermiculated with black.
Length, about twenty-six inches.

STONE CURLEW.

THE Stone Curlew (*Œdincnemus crepitans*, Temminck) is
a summer migrant to the British Islands, but examples are
occasionally obtained in winter.   It also occurs in temperate
Europe, Northern Africa, as well as South-Western Asia.
The adult has the upper parts light brown, streaked down
the centres of feathers with dark brown ; primaries, black,
the first two being banded with white ; between greater and
lesser wing-coverts a light band passes across ; feathers of
tail, faint brown, variegated with brown of a darker shade ;
a streak below the eye and throat, white ; neck and breast,
buff, with dark brown streaks ; abdomen, lighter ; vent and
under tail-coverts, whitish ; irides, rich yellow, and large ;
bill, light yellow at base, remainder black ; legs and feet,
yellow.   Length, about seventeen inches.   (P. 299.)

Stone Curlew.

Cream=coloured Courser.

THE Cream-coloured Courser (*Cursorius gallicus*, Gmelin) is a rare abnormal spring and autumn migrant to the British Isles, and is found inhabiting North Africa and South-Western Asia. The adult has the general colour of the plumage ochreous-buff, lighter on under parts; crown, buff, shading into dark grey on nape, and bordered with black; a band over the eye passing to the nape, white, and a black stripe below; primaries, black; throat, white; tail-feathers, marked with a black spot near the end; irides, hazel; bill, black, lighter at base beneath; legs and feet, dusky brown. Length, about ten inches. The male and female are alike in plumage.

Dotterel.

THE Dotterel (*Eudromias morinellus*, Linnæus) is a summer migrant to the British Isles, and is widely distributed through Northern and Central Europe and Western Asia, wintering in Africa north of the Equator. The adult female in summer has the upper parts ashy-brown, edged with paler; inner secondaries, margined with rufous; crown, blackish; eye-stripe, broad and white, sloping downward and running backward to the nape; tail, tipped with white, with the exception of rectrices; chin and upper throat, white; lower throat, light brown tinged with greyish, with some dark marks intervening; a white band passes across the chest, which is bordered on the upper side with black; lower breast and flanks, bright chestnut; abdomen, black; irides, hazel; bill, black; legs and feet, brownish-yellow. Length, about nine inches. The male is less than the female, and duller in colouration.

Common Pratincole.

THE Common Pratincole (*Glareola pratincola*, Linnæus) is a rare irregular spring and autumn migrant to the British Islands, and is found inhabiting the South-western Palearctic Region. The adult has the general colour of the plumage brownish-grey; throat, white, tinged with rusty, and banded by a narrow crescentic black band; lores, black; primaries and greater wing-coverts, blackish; tail-coverts and tail, white; the latter forked and banded broadly at the terminal end with brownish-black; breast, pale brownish-buff; below, white; axillaries, reddish-chestnut; irides, dark hazel; bill, reddish, black at the tip; legs and feet, dark brown. Length, from nine to ten inches. The adult female has the black on the lores less pronounced, and the crescentic band is absent.

Lapwing.

THE Lapwing (*Vanellus cristatus*, Wolf and Meyer) is a resident species in the British Islands, and is also an inhabitant of the Palearctic Region.   The adult male in nuptial plumage has the upper parts green, with glossy reflections of bluish-purple and reddish-purple; upper tail-coverts, bright chestnut; tail-feathers have the basal half white; rest, black, tipped with white; excepting outermost ones, which are nearly white; the crown and crest feathers, black, glossed with green; sides of neck, whitish; face, throat, and upper portion of breast, black, tinged with blue; abdomen, white; under tail-coverts, chestnut; irides, brown; bill, black; legs and feet, brownish-red.   Length, about thirteen inches.   The female is duller in colour than male, and crest shorter.   The young birds show no black on throat; crest also short.

Sociable Lapwing.

JOHN DUNCAN.

THE Sociable Lapwing (*Vanellus gregarius*, Pallas) is an
exceedingly rare straggling autumn migrant to the British
Isles, and is found inhabiting the South Central Palearctic
Region. The adult has the top of the head, lores, and a
streak behind the eye, black; a broad line of white runs
from the base of the bill over the eye to the nape; ear-
coverts, light buff; chin, white; hind neck, back, scapulars,
and wing-coverts, greyish; secondaries and tail, white;
central tail-feathers have a band of black near the terminal

end; primaries, mostly black; breast, ashy-grey; abdomen, black; flanks and vent, rich deep chestnut; under tail-coverts, white; irides, dark hazel; bill, legs, and feet, black. Length, about thirteen inches.

Killdeer Plover.

THE Killdeer Plover (*Ægialitis vocifera*, Linnæus) is a very rare abnormal autumn migrant to the British Islands, and inhabits the Nearctic Region with the exception of the extreme north. The adult of this plover can be distinguished from its British allies by the lower back, rump, and upper tail-coverts being rufous. Length, from nine to ten inches.

Greater Ringed Plover.

THE Greater Ringed Plover (*Ægialitis hiaticula major*, Tristam) is a common resident in the British Isles. It is also said to inhabit the Western Palearctic Region and the North-eastern Nearctic Region. The adult male in spring plumage has a band at the base of upper mandible, lores, crown, a patch below the eye, and a band across the chest, which narrows towards the nape, black; forehead, a stripe behind the eye, chin, throat, and a broad ring round the neck, white; upper parts mostly ashy-brown; primaries, brown; with white on the shafts forming a white bar across when the wing is extended; outer rectrices, chiefly white; remainder brown, tipped with white; lower breast and abdomen, white; irides, brown; bill, yellow at base, black at tip; legs and feet, orange. Length, about seven inches.

Ringed Plover.

THE Ringed Plover (*Ægialitis hiaticula*, Linnæus) is a common spring and autumn coasting migrant to the British Isles. It is also found inhabiting the Western Palearctic Region and North-eastern Nearctic Region.

Little Ringed Plover.

THE Little Ringed Plover (*Ægialitis minor*, Wolf and Meyer) is a rare irregular spring and autumn migrant to the British Isles, and inhabits the Palearctic Region. The plumage of the adult closely resembles that of the ringed plover, but can readily be distinguished by having the shafts of the quills all dusky with the exception of the outer one, which is white. Length, about six inches.

Kentish Sand Plover.

JOHN DUNCAN

THE Kentish Sand Plover (*Ægialophilus cantianus*, Latham)
is a summer visitor to the British Isles, and is also found
in the temperate parts of the Palearctic Region, making
its way in winter to Africa, India, and Southern China.
The adult has the upper parts pale brown, tinged with
greyish; secondaries, light brown; primaries, greyish-
brown, the former tipped with white; tail, hair-brown;
three outer feathers on each side, pale ochreous-white;
forehead, and a wide line passing over each eye, white;
the lores, black; a stripe behind the eye, black; fore part
of crown, black; top of head and nape, rich warm brown;
all round neck, and whole of under parts, white; neck on
each side has a black patch; irides, dark brown; bill, legs,
and feet, black. Length, about six inches. The female
has the plumage duller, no black on fore-crown; neck
spots, brown.

Caspian Sand Plover.

JOHN DUNCAN

THE Caspian Sand Plover (*Ægialophilus Asiaticus*, Pallas) is a very rare irregular spring migrant to the British Islands, and is found inhabiting the South Central Palearctic Region.

Golden Plover.

THE Golden Plover (*Charadrius pluvialis*, Linnæus) is a
resident species in the British Isles; it also occurs in
Northern Europe and the western portion of Siberia,
and is found during winter in Africa.   The adult in
summer has the upper parts nearly black, with copious
spots and bars of bright yellow and greyish; primaries,
black; bars on tail, brown; forehead, white; line above
eye, white; sides of neck, white; chin, throat, breast,
and belly, black, fringed with white on lower breast and
abdomen; axillaries and under tail-coverts, white; irides,
deep hazel; bill, legs, and feet, slaty-black.   Length, about
ten inches.   The female in same plumage resembles male,
but generally shows less black underneath.   The young
in first plumage have the breast suffused with yellowish,
and flanks profusely mottled with brown.

Grey Plover.

THE Grey Plover (*Charadrius Helveticus*, Brisson) is a common spring and autumn migrant to the British Islands, occasionally occurring in winter, and is found inhabiting the Northern Nearctic and Palearctic Regions. The adult male in summer has the general colour of the upper plumage whitish, barred with brown and black; forehead and eye-stripe, white; cheeks, lores, throat, breast, upper abdomen, and axillaries, black; thighs, vent, and under tail-coverts, white; tail, white, barred with brown and black; primaries, deep brown, with white marks on inner webs; irides, hazel; bill, legs, and feet, blackish. Length, from ten to twelve inches. The adult female in summer is duller in colour than the male.

Asiatic Golden Plover.

THE Asiatic Golden Plover (*Charadrius fulvus*, Gmelin) is an exceedingly rare abnormal autumn migrant to the British Isles, and is found inhabiting the North-eastern Palearctic Region.   This plover can be distinguished from *Charadrius pluvialis* by being smaller, and having grey axillaries, those of the latter bird being white.

American Golden Plover.

THE American Golden Plover (*Charadrius fulvus Americanus*, Schlegel) is a very rare irregular autumn migrant to the British Isles.   It also inhabits the Northern Nearctic Region.

COMMON STILT.

THE Common Stilt (*Himantopus melanopterus*, Meyer) is a rare irregular spring and autumn migrant to the British Isles, and inhabits the Southern Palearctic and Oriental Regions.   The adult male has the forehead, cheeks, front and sides of neck, lower portion of back, and under parts, white, with a blush of pink on the latter; occiput and nape of the neck, blackish; mantle, scapulars, wing-coverts, and primaries, black, tinged with green; tail, greyish; irides, crimson; bill, black; legs and feet, pink.   Length, about thirteen inches.   The adult female is said to have the back brownish.   (P. 313.)

Common Stilt.

Common Avocet.

THE Common Avocet (*Recurvirostra avocetta*, Linnæus) is a rare spring and autumn migrant to the British Isles, and inhabits the Southern Palearctic Region and Ethiopian Region. The adult has the cheeks and neck, and rest of plumage, white, with the exception of the head, nape, most part of hind neck, outer scapulars, middle wing-coverts, and primaries, which are black; irides, chestnut-brown; bill, black; legs and feet, pale blue. Length, about eighteen inches. The adult female resembles the adult male, but the colours are duller. The young have the black parts of the plumage washed with brown and margined with rusty.

COMMON CURLEW.

THE Common Curlew (*Numenius arquatus*, Linnæus) is a resident bird in the British Isles, moving about locally according to climatic conditions. It is found in summer in Northern Europe, and occurs during winter in Africa. The adult in nuptial dress has the prevailing colour of the upper parts pale brown; wing-coverts, dull white; centre of feathers, dark brown; primaries, blackish; lower back and rump, white, with bold spots of blackish; feathers of tail, dull white, barred with dark brown; lower parts, light brown; abdomen, whitish, and streaked with brown; flanks, streaked, shaded, and barred with brown; irides, hazel; bill, dark brown, paler at base; legs and feet, slate-grey. Length, about twenty inches. The female is larger than the male; the bill is also longer. The young in first plumage are very similar to adults, but show more rufous, and the markings beneath are more profuse. (P. 316.)

Common Curlew.

Whimbrel.

THE Whimbrel (*Numenius phæopus*, Linnæus) is a summer
visitor to the British Islands, and inhabits Northern
Europe.   The adult whimbrel in colouration of plumage
is similar to that of the curlew; but is much smaller
than the latter species, and the crown is dark brown
with a band of pale buffish-white down the centre;
eye-stripe, lightish; irides, hazel; bill, dark brown; base,
beneath, paler; legs and feet, slate-grey.   Length, about
seventeen inches.   The female, beyond being larger in
size, closely resembles the male.   The young in first
plumage have the upper parts marked with buff; streaks
on rump, dark brown.

Eskimo Whimbrel.

JOHN DUNCAN.

THE Eskimo Whimbrel (*Numenius borealis*, J. R. Forster) is an exceedingly rare straggling autumn migrant to the British Islands, and inhabits the Northern Nearctic Region. This whimbrel is easily distinguished from *Numenius phæopus* by its smaller size, nearly uniform coloured primaries, and the absence of white on the lower portion of back and rump. Irides, hazel-brown; bill, deep brown;

under mandible, paler at base; legs and feet, brownish-green.   Length, about fourteen inches.

Red=necked Phalarope.

THE Red-necked Phalarope (*Phalaropus hyperboreus*, Linnæus) is a summer migrant to the British Islands, but has not occurred in Ireland.   It is also met with in the Northern Palearctic Region and Nearctic Region.   The adult female in nuptial plumage has the top and sides of head, nape, hinder portion of neck, and most of upper parts, dull slate-grey, changing into brown on wings and rectrices; scapulars, marked with deep rufous; greater wing-coverts, broadly tipped with white ; innermost secondaries, narrowly tipped with white; upper tail-coverts have broad white bars; patch over eye and chin, white ; front of neck and sides, deep rufous; middle of breast and upper breast, slaty-grey ; rest of under parts, white; irides, brown; bill, black; legs and feet, greyish-olive.   Length, about seven inches.   The male is a little less in size and duller in colouration.

Grey Phalarope.

THE Grey Phalarope (*Phalaropus fulicarius*, Linnæus) is a rare nomadic spring, autumn, and winter migrant to the British Islands, and inhabits the Circumpolar Region. In winter it migrates down to the Equator. The adult female in summer has the head, nape, back, and scapulars, black-ish, broadly margined with rusty-red ; wing-coverts, blackish ; margined with white; white transverse bars on wings ; cheeks, white ; tail, dark grey, dusky towards terminal end ; front of the neck and under parts, chestnut ; irides, dark brown ; bill, pale orange ; legs, feet, and membranes, which are lobed, yellow. Length, about eight inches. The adult male is less in size, and in summer is duller in colour than the female. The adult in winter lacks the chestnut, the plumage being black, grey, and white.

Oyster Catcher.

THE Oyster Catcher (*Hæmatopus ostralegus*, Linnæus) is a resident species in the British Islands, and is met with in Northern Europe as well as Central Asia. The adult in summer has the head, neck, upper portion of breast, mantle, and scapulars, brilliant black; lesser wing-coverts, also black; tail, black; basal part of latter, white; greater wing-coverts, part of secondaries, rump, and upper tail-coverts, white; primaries, blackish, marked with white on inner webs; remainder of plumage, white; irides, crimson; bill, orange, getting richer in colour towards base; legs and feet, pinkish red. Length, about sixteen inches. The female resembles the male. The young have the upper parts margined with buffish, and show no white on throat.

Ruff.

THE Ruff (*Totanus pugnax*, Linnæus) is a rare summer migrant to the British Isles, and is also met with in the Northern Palearctic Region. It occurs in Africa and India during winter, and occasionally visits North-East America. This species varies so much in the colouration of the plumage that a detailed description would not be of very great service.   The adult male during the nesting season has the face covered with yellowish warty tubercules.   The colouration consists of brown of various shades—buffish, ochreous; white, black, glossy reflections of purple, green, etc.   By reference to the plate it will be seen that a ruff of feathers encircles the face and neck of the male.   Length, about ten or twelve inches.   The female, known as the reeve, is smaller than the male, and has not a ruff at any time.

Turnstone.

THE Turnstone (*Strepsilas interpres*, Linnæus), which inhabits the Circumpolar Region, is a common spring and autumn coasting migrant to the British Isles, and remains occasionally to winter.   The adult male in summer has the mantle and scapulars rich chestnut, varied irregularly with black; forehead, loral region, throat, nape, occiput, and hind neck, white; upper tail-coverts, white, marked with black; tail, white, with a black band near terminal end; crown, black, with whitish margins; patch below eye, and gorget on lower neck and upper breast, black; under parts, white; irides, dark brown; bill, blackish; legs and feet, rich orange.   Length, about nine inches.   The adult female

is duller in colour than the male.   The young in first
plumage lack the chestnut which is characteristic of the
adults in summer.

Bartram's Sandpiper.

JOHN DUNCAN.

BARTRAM'S Sandpiper (*Totanus Bartrami*, Wilson) is an
extremely rare abnormal autumn migrant to the British
Islands, and inhabits the Central Nearctic Region.   The
adult in summer has the ground colour of the upper
plumage rufous-brown, mottled and streaked with black
and white; head, neck, and breast, rusty; chin, abdomen,
and vent, white; fore-neck, breast, and flanks have arrow-
shaped markings of dark brown; inner web of first

primary, white, with black bars; axillaries, white, barred
with brown; tail, buffish-orange, reddish-brown, and white,
barred with black; irides, hazel; bill, yellowish-buff, darker
at tip; legs and feet, buffish.   Length, about twelve inches.

Curlew Sandpiper.

THE Curlew Sandpiper (*Tringa subarquata*, Guldenstadt)
is a common spring and autumn migrant to the British
Isles.   It is also met with in the Polar Regions.   The
adult in summer has the mantle, neck, and head, chestnut,
barred and streaked with grey and black; upper tail-coverts,
white, with a buffish tinge, and broadly barred with black;
primaries and tail, ashy-grey, with white shafts; beneath,
chestnut, shading on sides of body and vent to a rusty
colour, with markings of black; irides, hazel-brown; bill,

slightly decurved ; legs, feet, and claws, blackish.   Length, about seven inches.   The adult female is a little larger in size, and has a longer bill.

Spotteϸ Sanϸpiper.

THE Spotted Sandpiper (*Totanus macularius*, Linnæus) is an exceptionally rare straggling spring and autumn migrant to the British Isles, and inhabits the Nearctic Region. This sandpiper greatly resembles the common sandpiper, but the latter bird has the eighth and ninth secondaries nearly white, whilst the spotted sandpiper has all the secondaries broadly barred with ash-brown.

WOOD SANDPIPER.

THE Wood Sandpiper (*Totanus glareola*, J. F. Gmelin) is a rare visitor to the British Isles during migration, only one instance being recorded of its having nested in the British Islands.   It is also found in most parts of the Palearctic Region, and in autumn makes southward to Southern Africa, also India and the Philippine Isles. The adult differs from the green sandpiper in being smaller, having a shorter bill and wings and longer legs; upper parts darker, more olive and more mixed with white about the back ; outermost tail-feathers, white, barred or spotted with brown on outer webs; axillaries, white, obscurely barred with brown; irides, light brown ; bill, blackish ; legs and feet, pale dull olive. Length, about seven inches. The female resembles male.   (P. 327.)

Wood Sandpiper.

Green Sandpiper.

THE Green Sandpiper (*Totanus ochropus*, Linnæus) is a spring and autumn migrant to the British Islands, and is occasionally found in winter. It inhabits the Northern Palearctic Region. The adult in summer has the general colour of the upper plumage dusky olive-brown, with streaks of white on head and neck; mantle, scapulars, and innermost secondaries, spotted with white; primaries, brownish; rump, upper tail-coverts, and outer tail-feathers, white; remainder of rectrices, barred with blackish-brown; under parts, white, with streaks of brown on neck and chest; axillaries, brownish-black, with narrow bars of white; irides, dark hazel; bill, blackish; legs and feet, dark greyish, tinged at the points with green. Length, about eight inches. The female is like the male.

Common Sandpiper.

THE Common Sandpiper (*Totanus hypoleucus*, Linnæus) is a summer migrant to the British Isles, and also inhabits the Palearctic Region. The adult has the upper parts of a sandy brown, finely barred, streaked, and marked with zigzag lines of darker brown ; wings, brown tinged with blackish, and barred with white ; tail, brownish ; outer pairs of rectrices, barred with black and tipped with white ; eye-stripe, chin, and throat, white ; sides of neck and breast, suffused with pale brownish-greyish streaked with dark brown ; under parts and axillaries, white ; irides, hazel ; bill, blackish-brown ; legs and feet, olive, tinged with grey. Length, about seven inches. The female is similar to male. The young show buff margins on the upper feathers, and streaks on throat are absent.

Yellow=legged Sandpiper.

JOHN DUNCAN.

THE Yellow-legged Sandpiper (*Totanus flavipes*, Gmelin) is a very rare abnormal autumn migrant to the British Isles, and inhabits the Northern Nearctic Region.   The adult in nuptial plumage closely resembles the wood sandpiper in the same stage, but is larger in size, and the former shows less white on the rump in proportion, and there are indications of bars at all ages; bill, black and slender; legs and feet, clear yellow.   Length, about ten or eleven inches.

Solitary Sandpiper.

JOHN DUNCAN.

THE Solitary Sandpiper (*Totanus solitarius*, Wilson) is an extremely rare abnormal autumn migrant to the British Islands, and is found inhabiting North America.   The nearest ally to this sandpiper is *Totanus ochropus*, but it can be at once distinguished from the latter by the rump and upper tail-coverts, which are brown, sparingly spotted with white.   Length, about eight inches.

Bonaparte's Sandpiper.

JOHN DUNCAN.

BONAPARTE'S Sandpiper (*Tringa fusicollis*, Vieillot) is a rare straggling autumn migrant to the British Islands, and inhabits the Northern Nearctic Region.   The adult in summer bears a close resemblance to the dunlin, but can easily be distinguished from that bird at all stages, the upper tail-coverts being white, with a few longitudinal streaks of dusky.

purple Sandpiper.

THE Purple Sandpiper (*Tringa maritima*, Gmelin) is a
fairly common autumn migrant to the British Isles, and
inhabits the Northern Nearctic and Palearctic Regions.
The adult in summer has the upper parts of plumage
blackish, margined with rufous; some of the feathers have
creamy-white tips; wings, brownish, with a white band;
rump, tail-coverts, and two central rectrices, deep brown;
rest of rectrices, grey; superciliary stripe, whitish; under
parts, white, with greyish-brown centres to feathers on
breast and flanks; irides, warm brown; bill, deep brown;
base beneath, paler; legs and feet, dull yellow. Length,
about eight inches. The adult in winter has the upper
parts generally of a purplish-black, margined with grey;
head and neck, light greyish-black; breast and flanks, grey,
margined with white; centre of abdomen, white.

Broad=billed Sandpiper.

John Duncan

The Broad-billed Sandpiper (*Tringa platyrhyncha*, Tem-minck) is a rare irregular spring and autumn migrant to the British Isles, and inhabits the Northern Palearctic Region. The adult in summer has the general colour of the upper plumage blackish-brown, with chestnut margins to most of the feathers, with the exception of a few which are margined with white; primaries and central rectrices, blackish; outer tail-feathers, light ashy-brown; eye-stripe, neck, breast, flanks, and under tail-coverts, white, spotted with dark brown; remainder of under parts, white; irides, hazel; bill, which is deep brown and lighter beneath, is wide and flat and high at the upper base; legs and feet, blackish. Length, about six inches.

American Pectoral Sandpiper.

THE American Pectoral Sandpiper (*Tringa acuminata pec-toralis*, Say) is a rare straggling autumn migrant to the British Islands, and inhabits Arctic America, reaching Southern America in winter. The adult has the upper plumage deep brown, the feathers having rusty and cream coloured margins; secondaries, dusky-black, slightly margined with white; upper tail-coverts and two centre rectrices, blackish; remainder, greyish-brown, tipped with pale ochreous; face and throat, dullish white, with streaks of brownish; upper breast, whitish, tinged with brown, and closely streaked with blackish; under parts, white; irides, dark hazel-brown; bill, orange-brown at base; tip, dark brown; legs and feet, yellowish-brown. Length, about nine inches. The adult sexes are alike.

Siberian Pectoral Sandpiper.

THE Siberian Pectoral Sandpiper (*Tringa acuminata*, Hors-field) is a very rare abnormal autumn migrant to the British Isles, and inhabits North-Eastern Siberia.

Buff=breasted Sandpiper.

THE Buff-breasted Sandpiper (*Tringa rufescens*, Vieillot) is an exceedingly rare straggling autumn migrant to the British Islands, and inhabits the Northern Nearctic Region. This sandpiper can readily be recognised by the under sides of the primaries and secondaries, which are pure white, beautifully mottled with dark brown : the tail is cuneiform,

and the centre rectrices black; remainder, light brown,
enclosed by a zone of black, and edged with white.
Length, about eight inches.   The adult sexes are alike in
colouration.

Little Stint.

THE Little Stint (*Tringa minuta*, Leisler) is a fairly common
coasting migrant in spring and autumn, and inhabits the
North-western Palearctic Region.   The adult in summer
has the general colour of the upper parts blackish-brown,
with chestnut margins on the feathers; the chestnut on
the upper tail-coverts is, however, scarcely discernible;
primaries, brown; greater wing-coverts, tipped with white,
forming a white bar across the wings; two central rectrices,
blackish-brown, narrowly margined with chestnut; rest of
tail-feathers, grey; under parts, white, tinged with buff on
the breast, the latter being faintly spotted with brown;
irides, hazel; bill, legs, and feet, black.   Length, between
five and six inches.

22

American Stint.

JOHN DUNCAN

THE American Stint (*Tringa subminuta minutilla*, Vieillot) is an exceedingly rare straggling autumn migrant to the British Isles, and inhabits the Northern Nearctic Region. This stint closely resembles the little stint in colouration of plumage, and is considered by some naturalists to be merely a local race of *Tringa minuta*.

Temminck's Stint.

TEMMINCK'S Stint (*Tringa Temmincki*, Leisler) is a rare spring and autumn migrant to the British Isles, and inhabits the Northern Palearctic Region. The adult in nuptial plumage has, as a rule, the general colour of the upper parts greyish-brown, several of the feathers more or less deep brown, barred and edged irregularly with creamy chestnut; axillaries, white; primaries, brown; greater wing-coverts, tipped with white; four central rectrices, deep brown; remainder, white; throat and breast, creamy brown, streaked with brownish; under parts, white; irides, hazel; bill, blackish; legs and feet, greyish-brown. Length, about five inches.

Sanderling.

THE Sanderling (*Tringa arenaria*, Linnæus) is a common coasting autumn migrant to the British Isles, and inhabits the Northern Nearctic and Palearctic Regions. The adult in summer plumage has the feathers of back and scapulars, rufous, with patches and spots of black; primaries, blackish; greater wing-coverts, blackish-brown, tipped and margined with white; forehead and crown, black, with light rufous and white margins to the feathers; cheeks, neck, and upper breast, light chestnut, spotted with deep brown; rest of under parts, white; irides, hazel; bill, legs, and feet, black. Length, between seven and eight inches. The sanderling has no hind toe. The winter plumage is greyish on the upper parts and white below.

Dunlin.

THE Dunlin (*Tringa alpina*, Linnæus) is a resident species in this country, and is also found inhabiting the Palearctic Region and some portions of the Nearctic Region.   The adult in summer has the feathers on the crown of the head black, with chestnut margins; chin, white; fore-part of neck and breast, greyish-white, streaked with dark brown; abdomen, black; remainder of under parts, white; hind part of neck, mantle, and scapulars, black, broadly margined with chestnut; greater coverts, with white tips; wings and tail, greyish-brown, central pair of rectrices, darker; irides, hazel; bill, legs, and feet, black.   Length, from six to seven inches.   The female is the same as the male in colouration, but a trifle larger in size.

Knot.

THE Knot (*Tringa canutus*, Linnæus) is a fairly common migrant to our shores in spring and autumn.   It is supposed to breed on all suitable places in the North Polar Basin. The adult in summer has the feathers of the mantle variegated with black, chestnut, and white ; crown and neck, reddish-brown, streaked with darker ; tail-coverts, white with bars of black ; sides of head, throat, and breast, chestnut ; remainder of under parts, whitish, with black mottlings ; irides, brown ; bill, legs, and feet, blackish.   Length, about nine inches.

Redshank.

THE Redshank (*Totanus calidris*, Linnæus) is a resident in
the British Isles.   It also inhabits the Palearctic Region,
but wends its way to Africa to pass the winter.   The adult
in summer is pale brown on the upper parts, streaked
profusely and barred with umber; secondaries, nearly
white; rump, white flecked with dusky; rectrices, white,
with blackish bars, central pairs tinged with light brown;
under parts, white, streaked on the neck and breast, and

with bars of umber on flanks ; irides, hazel ; bill, orange ;
tip, black ; legs and feet, orange-red.   Length, about eleven
inches.

Dusky Redshank.

THE Dusky Redshank (*Totanus fuscus*, Linnæus) is a rare
spring and autumn migrant to the British Islands, and also
occurs occasionally in winter.   It inhabits the Northern
Palearctic Region.   The adult in summer has the upper
parts of body blackish-brown, with white triangular spots

upon tips and margins ; the head and neck, brownish-black, tinged with grey; primaries, dusky black ; rump and tail, white, the latter barred closely with blackish ; under parts, brownish-black, washed with grey; the feathers on breast and abdomen, edged finely with white; irides, hazel ; bill, black ; base beneath, orange-red; legs and feet, orange, tinged with brown.   Length, about twelve inches.   The adult female shows more white on the under plumage than the male.

Greenshank.

THE Greenshank (*Totanus glottis*, Linnæus) is a summer migrant to the British Islands, although it is said that a few remain in Ireland during the winter.   It is also found inhabiting the Northern Palearctic Region.   The adult in summer has the mantle and secondaries black, some of the feathers with greyish margins ; quills, blackish ; rump, white; tail, white, barred and mottled with brownish; head and neck, white, tinged with grey and streaked with blackish-brown; wing-coverts, deep brown; breast, upper part of belly and flanks, white, spotted and streaked with greyish-brown; irides, hazel; bill, black; legs and feet, olive. Length, about thirteen inches. The female has the upper parts less conspicuously marked than the male. The young have the feathers on back margined with tawny, and chest and flanks pencilled finely with dark grey.

Black=tailed Godwit.

The Black-tailed Godwit (*Limosa melanura*, Leisler) is one of our regular spring and autumn coasting migrants to the British Islands, and is found inhabiting the Western Pale-arctic Region. The adult male in nuptial garb has the crown reddish-brown, with streaks of black; forehead, superciliary stripe, and chin, whitish; cheeks, throat, and breast, chestnut, the latter barred with blackish; back and scapulars, blackish-brown; feathers, barred and margined with chestnut; wings, deep brown, with a plainly defined white bar; tail, white at base, black at terminal end; abdomen, whitish, with bars of dark brown; irides, hazel-brown; bill, deep brown, orange at base; legs and feet, black.   Length, between eighteen and nineteen inches. The adult female in summer is larger than the male, and the colouration is much duller.

Bar=tailed Godwit.

THE Bar-tailed Godwit (*Limosa rufa*, Brisson) is a common spring and autumn migrant to the British Isles, occurring also in winter.   It inhabits the Northern and Western

Palæarctic Regions. The adult male in summer can readily be distinguished from the Black-tailed Godwit by the tail-feathers, which are whitish and barred with brown. The adult female, which is larger than the male, also has the bars on the tail. Length, about fifteen inches. The young in first plumage have the tail barred.

Woodcock.

THE Woodcock (*Scolopax rusticola*, Linnæus) is a partial resident, nesting sparingly in various parts of the British Isles; it also breeds in the temperate portions of the Palæ-arctic Region. The adult has the fore part of crown ashy-brown; hinder part and nape, pale chestnut, broadly barred with black bands; remainder of upper parts variegated with rufous, ash, grey, and brown; chin, white; general colour of under parts, greyish-buff, barred with brown; irides, deep brown; bill, legs, and feet, dull flesh colour. Length, about fourteen inches. The adult female resembles the male. The young also closely resemble the adults.

Great Snipe.

THE Great Snipe (*Scolopax major*, Gmelin) is a rare abnormal spring and autumn migrant to the British Isles, and inhabits the Western Palearctic Region.   The adult bears a close resemblance to the common snipe in colouration of plumage, but can always be distinguished from that bird by the number of tail-feathers, which in the latter are fourteen, whilst in *Scolopax major* the number is sixteen.   The great snipe is also a little larger, measuring from eleven to twelve inches.   The young in first plumage closely resemble adults.

Common Snipe.

THE Common Snipe (*Scolopax gallinago*, Linnæus) is a
resident in the British Isles, and is also found inhabiting
the Palearctic Region.   The adult has the crown of head
dark brown, divided by a light ochreous streak running
through the centre; lores, dark brown; eye-stripe, pale
yellowish; remainder of upper parts, cream-yellow, barred,
mottled, and otherwise marked with white and shades of
brown; secondaries, tipped with white; chin, white; throat
and breast, light brown, with markings of dusky-brown;
abdomen, white; sides, barred with dusky; tail consists

of fourteen feathers; irides, hazel; bill, light brown, darker at tip; legs and feet, brownish. Length, between ten and eleven inches. The female is similar to the male in colouration. The young show more uniformity in colouration of the upper parts.

Jack Snipe.

THE Jack Snipe (*Scolopax gallinula*, Linnæus) is a common autumn and winter migrant to the British Islands, and inhabits the Northern Palearctic Region. The adult in colouration of plumage is very similar to the common snipe, but it is smaller in size, and has only twelve feathers in the tail. Hence it is easily distinguishable from *Scolopax gallinago*. Length, about eight inches. The adults of both sexes are alike in colouration.

Red=breasted Snipe.

JOHN DUNCAN

THE Red-breasted Snipe (*Ereunetes griseus*, Gmelin) is a rare straggling autumn migrant to the British Islands, and inhabits the North Central Nearctic Region. The adult in summer has the upper back and scapulars black, the feathers margined and varied with light reddish-brown and white; lower back, rump, and rectrices white, with transverse bars of black; crown, blackish-brown, margined with light reddish-brown; lores, brownish; superciliary stripe and chin, white, suffused with chestnut-buff; shaft of first primary feathers, white; front neck, breast, and flanks, light buffish-chestnut, spotted and barred with deep brown; centre of abdomen, whitish; irides, hazel; bill, deep brown; legs and feet, olive-brown. Length, about ten inches.

Great Skua.

JOHN DUNCAN

THE Great Skua (*Stercorarius catarrhactes*, Linnæus) is only known, whilst resident with us, to nest in the Shetlands, and is mostly met with on its way to and from more northern districts during autumn and winter. It also occurs in North-West Europe, and in the eastern parts of Arctic America. The adult has the general colouration of the plumage dark brown, with rufous or greyish margins to the feathers ; basal portion of quills, white ; irides, hazel-brown; bill, black ; legs and feet, slate-black. Length, from twenty-two to twenty-five inches. The female is similar to the male. The young are said to be more uniform in colour than the adults.

23

𝕭uffon's 𝕾kua.

JOHN DUNCAN

BUFFON'S Skua (*Stercorarius Buffoni*, Boie) is a rather rare
nomadic autumn and spring migrant to the British Islands,
and inhabits the Northern Nearctic and Palearctic Regions.
The adult has the upper plumage mostly brownish-grey;
crown and nape, blackish; throat, white, tinged with yellow,
the latter colour extending to the sides of the neck and
across the lower ear-coverts to the nape; primaries, dark
brown; central rectrices (much elongated, narrow, and
pointed), blackish; under parts, mostly white; flanks and
under tail-coverts, brownish-grey; irides, brown; legs,
yellowish-grey; feet, black; bill, deep horn colour.   Length,
from tip of bill to end of longest tail-feathers, about twenty-
three inches.

Richardson's Skua.

JOHN DUNCAN.

RICHARDSON'S Skua (*Stercorarius Richardsoni*, Swainson) is one of our summer migrants. It inhabits both the Northern Palearctic Region and the Nearctic Region. The adult has the upper parts mostly of a slaty-grey; neck, tinged with yellowish-buff; beneath, white; sides of belly, lower portion of latter, and under tail-coverts, tinged with brownish-grey; central tail-feathers, elongated and reaching considerably beyond the ends of wing quills, the latter having white shafts at all ages; irides, hazel; bill, greyish; tip, black; legs and feet, black. Length, from twenty to twenty-one inches. The female is similar to the male, but the two central rectrices are not so long. The young after the first moult have the basal half of feet yellow.

Pomatorhine Skua.

JOHN DUNCAN.

THE Pomatorhine Skua (*Stercorarius pomatorhinus,* Temminck) is a more or less common autumn and spring
coasting migrant to the British Isles, examples occasionally being met with in winter.   It inhabits the Northern
Nearctic and Palearctic Regions.   This skua varies
considerably in the colouration of the plumage.   There
is a light and dark variety, neither of which appears to
be influenced by sex or age.   The adult of the light form
has the upper plumage mostly dark brown; the two central
rectrices reaching about four inches beyond the others, and
twisted upwards; crown, black; neck, white, tinged with
buffish-yellow; breast, dullish white; lower belly and under

tail-coverts, brown; irides, brown; bill, deep horn colour;
tip, black; legs and feet, black.   Length, about twenty-one
inches.

Kittiwake.

THE Kittiwake (*Larus tridactylus*, Linnæus) is a resident
in the British Isles.   It is also met with in the Northern
Palearctic Region, as well as the North-east Nearctic
Region.   The adult in nuptial dress has the mantle and
wings pale bluish-grey; head, neck, under parts, and tail,
white; secondaries, with white tips; outer primaries on
terminal portions, black; irides, hazel; bill, greenish-yellow;
legs and feet, brownish dusky-green; hind toe absent at
all times.   Length, from fifteen to sixteen inches.   The
female is similar to the male.   The young in first plumage
have the crown and nape deep grey, and the feathers on
the back are margined with brown.

Herring Gull.

JOHN DUNCAN

THE Herring Gull (*Larus argentatus*, Gmelin) is a resident in the British Islands, and is also found on the Northern and Western European coasts, and upon the eastern portions of the Nearctic shores. The adult in summer has the mantle, scapulars, and wing-coverts, light French grey; secondaries, tipped with white; primaries, mostly black on outer webs of the first two; inner webs, grey on the half towards the base; the first with a rounded triangular-shaped mark of white at the end; second and third, similar, but less in size; remainder of plumage, snowy white; irides, pale yellow; bill, bright yellow; angle of lower mandible, red; legs and feet, flesh colour. Length, from twenty-one to twenty-five inches. The female is similar to the male. The young in first plumage have the general colour composed of white, greyish-white, and browns of various shades.

GREAT BLACK-BACKED GULL.

THE Great Black-backed Gull (*Larus marinus*, Linnæus) is a resident species, which is more widely dispersed throughout the British Isles in winter than during the breeding season. It is also met with in most parts of Northern Europe, as well as North-Eastern America. The adult in summer is similar in colouration to the Lesser Black-backed Gull, but its larger size and flesh-coloured feet at once distinguishes from the latter species. Length, about thirty inches. The female is similar in colouration to the male, but is a little less in size. The young have the colours of plumage composed of greyish-white, greyish-brown, and buffish, similar to *Larus fuscus* and *Larus argentatus*; but the larger size is sufficient to define the species. (P. 360.)

Great Black backed Gull.

Lesser Black=backed Gull.

JOHN DUNCAN

THE Lesser Black-backed Gull (*Larus fuscus*, Linnæus) .
is a resident in the British Isles, but subject to much local
movement after the nesting season.   This gull is also found
inhabiting most parts of Europe and North Africa.   The
adult in summer has the mantle and wings dark slate;
innermost secondaries and longest scapulars, with white
tips; quills black, tipped with white; remainder of plumage,
pure white; irides, light yellow; bill, yellow, angle of lower
mandible, red; legs and feet, rich yellow.   Length, from
eighteen to twenty-three inches.   The female is a little
less than the male.   The young are similar to immature
examples of *Larus argentatus*, but the general colouration
of the upper parts is darker.

Common Gull.

John Duncan

THE Common Gull (*Larus canus*, Linnæus) is one of our
resident gulls in the British Isles, but moves about a great
deal during the non-nesting season.   It also occurs in the
Northern Palearctic Region.   The adult in breeding dress
has the back, wings, and coverts, pearl-grey; secondaries,
broadly bordered and tipped with white; primaries, mostly
black, with white spots or tips; remainder of plumage,
pure white; irides, hazel; bill, yellow at tip; base,

greenish; legs and feet, greenish-yellow. Length, about eighteen inches. The female is similar to the male, but is a little less in size. The young have the upper parts more or less marked with greyish-brown, dull brown, and greyish; primaries, dusky-brown; tail at terminal end, banded with dusky-brown.

Black=headed Gull.

THE Black-headed Gull (*Larus ridibundus*, Linnæus) is another resident species in the British Islands, and, like many other of our resident gulls, is subject to much local, also southern movement after the nesting season. It is also met with in the temperate parts of Europe and Asia. The adult in summer plumage has the mantle and wing-coverts pearl-grey; quills, marked with white in centre, and both webs margined with blackish; head and upper part of neck, dark brown; rest of plumage, white; irides, hazel; bill, feet, and legs, coral-red. Length, about sixteen inches. The female is similar to the male. The young have the upper parts more or less brownish, with paler tips to the feathers; tail, at terminal end, broadly banded with blackish.

Great Black=headed Gull.

JOHN DUNCAN.

GREAT Black-headed Gull (*Larus ichthyaetus*, Pallas) is an exceedingly rare abnormal spring migrant to the British Isles. It inhabits the South Central Palearctic Region and North-eastern Ethiopian Region. The adult in breeding plumage is described as having the head deep black ; mantle

and wing-coverts, dark grey; secondaries, tipped broadly
with white; first quill, white, with a narrow band of black
on outer web and a patch of black on inner web; next
three primaries, with subterminal bars of black and white
tips; remainder of plumage, white; a small white patch
behind the eye; irides, deep brown; bill, rich yellow, black
and red at the angle; legs and feet, greenish-yellow; webs,
orange.   Length of male, about twenty-seven inches.

Mediterranean Black=headed Gull.

JOHN DUNCAN

THE Mediterranean Black-headed Gull (*Larus melano-
cephalus*, Natterer) is an exceptionally rare abnormal winter
migrant to the British Isles, and inhabits the South-western
Palearctic Region.   The adult in summer has the head
black; secondaries and wing-coverts, pearl grey; primaries,
white, excepting the first quill, which has a black line on
outer web; rest of plumage, white; irides, brown; bill,

coral-red, and with a dark band in front of the angle; legs, brownish. Length, about seventeen inches. The adult in winter has the head streaked with dark brown.

Bonaparte's Gull.

BONAPARTE'S Gull (*Larus Philadelphia*, Ord) is an extremely rare straggling winter and spring migrant to the British Islands, and is found inhabiting the Northern Nearctic Region. The adult in summer has the mantle pearl-grey; hood, greyish-black; primary coverts, white; primaries, tipped with white; plumage otherwise similar to *Larus ridibundus*; bill, black. Length, about fourteen inches. The adult in winter is said to have the head nearly white.

Little Gull.

THE Little Gull (*Larus minutus*, Pallas) is a comparatively rare autumn and winter migrant to the British Islands, and inhabits the Northern Palearctic Region. The adult in summer has the whole head black; back, wings, and coverts, pearl-grey; axillaries, greyish; under parts of wings, brownish-black; primaries, tipped broadly with white; remainder of plumage, white, with a rosy flush on breast; irides, hazel-brown; bill, deep red; legs and feet, bright red. Length, from ten to eleven inches. The adults in winter have the head white, with marks of grey on the nape. The young in first plumage are more or less marked with blackish, grey, and white; tail, white; terminal end with a broad band of black.

Sabine's Gull.

JOHN DUNCAN —

SABINE'S Gull (*Xema Sabini*, J. Sabine) is a rare nomadic autumn migrant to the British Isles, and inhabits the Northern Nearctic and Palearctic Regions. The adult in summer has the head and nape dark slate; a narrow black collar round the neck; mantle, scapulars, and most of wing-coverts, slate-grey; primaries, black, with white on the outer half of the inner webs, and tipped with white; tail, white, and forked; rest of plumage, white; irides, dark hazel; bill, black at base; bright red at tip; legs and feet, slaty-black. Length, about fourteen or fifteen inches. The young in first plumage are more or less greyish, white, and black; tail, tipped with black.

Ross's Gull.

Ross's Gull (*Rhodostethia rosea*, Macgillivray) is an exceptionally rare straggling winter migrant to the British Isles, and inhabits the Polar Regions. The adult in summer plumage has the head and neck nearly white, the latter encircled with a narrow black collar; remainder of upper plumage, pearl-grey, changing into white on the tips of greater wing-coverts and secondaries; outer web of first quill feather, black; tail, white and cuneated; under parts, white, tinged with rose colour on breast and abdomen; irides, hazel; bill, black; legs and feet, bright red. Length, about thirteen inches. The adults in winter lack the black collar.

Glaucous Gull.

JOHN DUNCAN

THE Glaucous Gull (*Larus glaucus*, Fabricius) is an irregular straggling winter migrant to the British Isles, and is
found inhabiting the Northern Nearctic and Palearctic
Regions.   The adult does not differ to any great extent
from the Iceland Gull, but it is larger in size, and the pro-
portionably shorter wings only reach a little beyond the
tail, whilst those of the Iceland Gull extend considerably
beyond the end of the tail-feathers.   Length, about thirty-
two inches.

Iceland Gull.

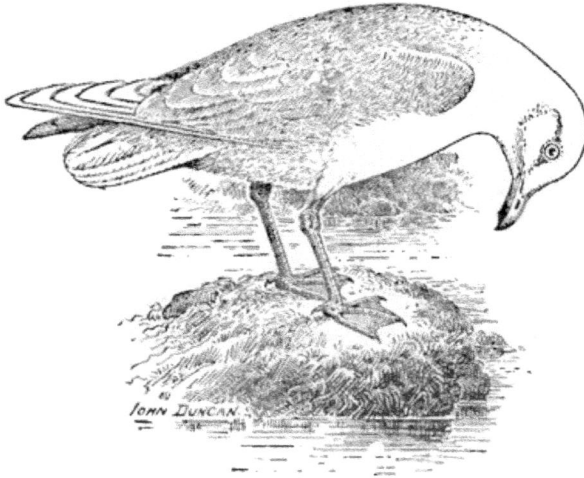

THE Iceland Gull (*Larus leucopterus*, Faber) is a rare straggling autumn and winter migrant to the British Isles, and is found inhabiting the Northern Nearctic Region. The adult in summer has the mantle and wing-coverts pale grey; remainder of plumage, white; irides, light yellow; bill, yellow, red at the angle beneath; legs and feet, pinkish flesh tint.   Length, about twenty-two inches.   The adult in winter has the head and neck streaked with grey.   The young have the upper plumage creamy-white, more or less varied with light brown markings.

Ivory Gull.

JOHN DUNCAN

THE Ivory Gull (*Pagophila eburnea*, Phipps) is a rare nomadic autumn and winter migrant to the British Islands,
and is found inhabiting the Circumpolar Regions. The
adult has the whole of the plumage pure white; irides,
deep brown; bill, greenish-grey, tip, yellow; legs and feet,
black. Length, from sixteen to eighteen inches. The
young in first plumage are described as being nearly of
a uniform ashy-grey.

Sandwich Tern.

THE Sandwich Tern (*Sterna cantiaca*, Gmelin) is a summer
visitor to the British Isles, and is also found inhabiting
most parts of temperate Europe and Northern Africa, as well
as South-Western Asia. The adult in summer has the forehead, crown, and occipital feathers, rich black; back and
wings, pearl-grey; tips of secondaries and upper quills, nearly
white; fore part of breast and under parts, tinged with
delicate pink; upper and under tail-coverts and tail, white
and forked; bill, black, tipped with pale yellow; legs and
feet, black. Length, about fifteen inches. The female is
shorter in the tail, and shows less pink on breast. The
young are pale brown above, varied with black; primaries,
greyish-black, margined and tipped with white.

Roseate Tern.

THE Roseate Tern (*Sterna Dougalli*, Montagu) was at one time a fairly numerous summer visitor to the British Islands. This tern is found in most parts of the temperate and tropical regions of the two hemispheres. The adult in summer plumage has the back and wings pearly-grey; upper tail-coverts, tinged with grey; outer web of first primary feather, black; remainder, greyish; whole of inner webs with a streak of greyish next the shaft; other portions of quills, white; tail, greyish-white and forked, the outside feathers reaching about three inches beyond primaries; forehead, crown, and occiput, rich black; remainder of plumage, white, tinged on breast with a rosy hue; irides, dark hazel; bill, red, towards tip, black; legs and feet, red. Length, from fifteen to seventeen inches. The female resembles the male.

Common Tern.

THE Common Tern (*Sterna hirundo*, Linnæus) is one of
our summer visitors to the British Isles, and is found in
most parts of the Palearctic Region, as well as the eastern
portions of North America. The adult in nuptial plumage
has the mantle and wings pearl-grey; crown and nape,
black; rump and tail, white, longest tail-feathers, grey on
outer margins; under parts, greyish on breast and belly;
irides, dark brown; bill and legs, deep orange red, blending
on the former into blackish towards the tip. Length, from
thirteen to fifteen inches. The female is similar to the
male. The young have the crown and nape mottled with
brownish; the mantle is mottled with greyish. This tern
can readily be distinguished from *Sterna Arctica* by its
longer tarsi.

Arctic Tern.

THE Arctic Tern (*Sterna Arctica*, Temminck) is a summer migrant to the British Islands. It is also found nesting in the Circumpolar Region, moving in autumn southward along the coasts of the Atlantic to Africa. The adult in summer resembles to a great extent *Sterna hirundo* in colouration of plumage, but the upper parts are darker, and the tarsi are much shorter. The bill is uniform coral-red. Length, about fifteen inches. The female is a shade less in size.

Lesser Tern.

THE Lesser Tern (*Sterna minuta*, Linnæus) is one of our late summer migrants to the British Islands. It is also an inhabitant of the Palearctic Region, ranging eastwards to North India. The adult in summer plumage has the back and wings light grey; quills, deeper in colouration, and having white margins on inner webs; forehead, white; rump and tail, white; crown of head, lores, and nape, black; remainder of plumage, white; irides, dark brown; bill, black at tip, deep yellow at base; legs and feet, orange. Length, from eight to nine inches. The female is similar to the male. The young have the upper parts mottled with blackish-brown; bill, legs, and feet, brownish.

Caspian Tern.

JOHN DUNCAN

THE Caspian Tern (*Sterna Caspia*, Pallas) is a rare abnormal spring and autumn migrant to the British Isles, and
is found inhabiting the temperate and tropical zones. The
adult in summer has the crown and nape of the neck rich
black; upper parts and tail, which is slightly forked, pearl-
grey; primaries, brownish, washed with grey; sides of the
head, front of neck, and rest of under parts, white; irides,
dark hazel; bill, vermilion; legs and feet, black. Length,
from nineteen to twenty-one inches. The forehead and
crown of the adult in winter is white, streaked with black;
behind the ear-coverts there is a dark patch. The young
in first plumage have ashy-brown mottlings on the upper
parts; forehead and crown, white; under parts, white.

Gull-billed Tern.

JOHN DUNCAN.

THE Gull-billed Tern (*Sterna anglica*, Montagu) is a rare straggling spring and autumn migrant to the British Isles. It inhabits the Southern Palearctic and Nearctic Regions. The adult male in nesting plumage has the forehead, crown, and back part of the neck, rich deep black; upper parts, pearl-grey; primaries, greyish; under parts, white; irides, brown; bill, stout and black; legs and feet, black. Length, from fourteen to fifteen inches. The adult in winter has the forehead and crown white, with grey streaks; ear-coverts, patched with grey. The young on upper parts are more or less marked with various shades of brown; under parts, white; legs and feet, brownish.

Sooty Tern.

Jᴏʜɴ Dᴜɴᴄᴀɴ

Tʜᴇ Sooty Tern (*Sterna fuliginosa*, Gmelin) is a very rare straggling autumn migrant to the British Islands, and is found inhabiting the Oceanic tropical zone.   The adult has the crown, lores, and nape, black; forehead and sides of the neck, white; back, scapulars, wings, and remainder of upper plumage, mostly black; outer webs of the two elongated outer rectrices, white; under parts, white; irides, hazel; bill, legs, and feet, blackish.   Length, about sixteen inches. The young are darker on the upper parts, with lightish tips to most of the feathers; under parts, brownish.

Noddy Tern.

JOHN DUNCAN

THE Noddy Tern (*Anous stolidus*, Linnæus) is an exceedingly rare irregular autumn migrant to the British Islands,
and is found inhabiting the Oceanic tropical zone. The
adult has the general colouration of the plumage dark-
brown, with a patch of grey on the forehead and crown;
the tail is graduated. Length, about sixteen inches.

Whiskered Tern.

THE Whiskered Tern (*Hydrochelidon hybrida*, Pallas) is an exceedingly rare abnormal spring and autumn migrant to the British Isles. It inhabits the Southern Palearctic and Oriental Regions, wintering in South Africa, and eastward has occurred as far south as Australia. The adult in breeding plumage has the head and nape of neck black; upper parts, lightish-grey; sides of face, chin, throat, and tail, white; under wing-coverts, white; chest, light grey; belly and flanks, blackish; irides, dark hazel; bill, deep red; legs and feet, bright red. Length, about eleven inches. The adult in winter has the forehead white; crown, mottled with white; under plumage, white. The young in first plumage have the feathers of upper parts margined with reddish.

Black Tern.

THE Black Tern (*Hydrochelidon nigra*, Linnæus) is a fairly
regular spring and autumn coasting migrant to the British
Isles, and inhabits the South-western Palearctic Region.
The adult in summer has the head and neck deep slate-
grey ; crown and nape, blackish; under wing-coverts, light
grey; upper parts and tail, dark bluish-grey, the latter
forked ; throat, breast, and abdomen, deep slate-grey ; vent
and under tail-coverts, white ; irides, dark hazel ; bill, black;
legs and feet, deep reddish-brown.   Length, about ten
inches.   The adult in winter has the crown, nape, forehead, cheeks, throat, and front neck, white ; remainder of
plumage, as in summer.

White=winged Black Tern.

THE White-winged Black Tern (*Hydrochelidon leucoptera*, Meisner and Schinz) is a rare irregular spring and autumn migrant to the British Isles, and inhabits the Southern Palearctic Region. The adult in summer has the head, crown, neck, nape, and back, rich shining black; lesser wing-coverts and shoulders, white; greater wing-coverts and secondaries, grey; axillaries, black; upper tail-coverts and tail, white; quills, blackish, tinged with grey; under parts, black; vent, white; irides, dark hazel; bill, red; legs and feet, reddish-orange. Length, about nine inches. After the autumn moult the head and neck of this tern are white, marked on the nape with black; mantle, slate-grey; under wing-coverts, greyish; rest of under parts, white.

Common Guillemot.

THE Common Guillemot (*Uria troile*, Linnæus) is a well-known resident in the British Islands, and is also met with on the northern coasts of Europe and the North Atlantic. The adult in nuptial plumage has the head, neck, chin, throat, and remainder of upper parts, blackish-brown, sometimes tinged with greyish; secondaries, tipped with white; under parts, white; irides, brown; bill, blackish; legs and feet, black; webs of an olive tinge. Length, about eighteen inches. The female is the same in colour as the male, but is a trifle less in size. The young have shorter bills, and the legs are paler.

Ringed Guillemot.

THE Ringed Guillemot (*Uria lacrymans*, Macgillivray) is generally conceded to be merely a variety of the common guillemot. It is readily distinguished by having a white ring round the eye, and also a white streak behind it.

Brunnich's Guillemot.

BRUNNICH'S Guillemot (*Uria Brunnichi*, Sabine) is an extremely rare nomadic autumn migrant to the British Islands, and is found inhabiting the Circumpolar Region. The adult closely resembles the common guillemot in the colouration of the plumage, but can be distinguished from the latter by the short and stout bill, which has horn-coloured lines on the upper mandible, reaching on each side from the gape to the nostrils.

Black Guillemot.

THE Black Guillemot (*Uria grylle*, Linnæus) is a resident
in the British Isles, but moves about a great deal after nesting time.   It also frequents the islands and coasts of the
North Atlantic.   The adult male in summer has the whole
of the plumage black, with the exception of a large patch of
white on the wing-coverts ; irides, hazel ; bill, black ; legs
and feet, bright red.   Length, from twelve to fourteen
inches.   The female is similar to the male.   The young
have the wing-patch marked with brown, and show brown
on the head and neck ; legs and feet, dark brown.

𝕷ittle 𝕬uk.

THE Little Auk (*Mergulus alle*, Linnæus) is an irregular straggling autumn and winter migrant to the British Islands, and inhabits the North Atlantic and Arctic Ocean basins. The adult in summer has the head, chin, throat, and upper parts, mostly glossy black; a small white spot over each eye; some scapulars margined, and secondaries tipped, with white; under parts, below upper breast, also white; irides, hazel; bill, blackish; legs and feet, fleshy-brown; toes, lighter. Length, about eight inches. The winter plumage of the adult is similar to the summer plumage, except that the throat and chest are white. The young lack the white

spots over the eyes, and the white on the scapulars, and the
black parts are more sooty.

Puffin.

JOHN DUNCAN.

THE Puffin (*Fratercula Arctica*, Linnæus) is a resident in
the British Isles, and it also inhabits the coasts and islands
of the North Atlantic.   The adult male in summer has the
forehead, crown, occiput, upper parts, and a broad collar
round the neck, black ; primaries, dark brown ; cheeks and
chin, white, tinged with ashy-grey ; remainder of plumage,

white; irides, greyish; bill, pale slate at base; centre, yellow; tip, orange-red; legs and feet, orange. Length, from eleven to twelve inches. The female is similar to the male, but the bill is not so much developed. The young are duller in colouration, and show more grey on cheeks.

Razorbill.

JOHN DUNCAN.

THE Razorbill (*Alca torda*, Linnæus) is a widely and commonly distributed resident species in the British Isles, and also inhabits most of the Palearctic and also the

Nearctic Regions. The adult male in breeding plumage
has the whole of the upper parts black, tinged in places with
green; secondaries, tipped with white; a pure white line
runs from the eye to the top of the bill; remainder of
plumage, white; irides, dark hazel; bill, black, with a white
band across the centre on both sides; legs and feet, blackish.
Length, about seventeen inches. The female is like the
male. In the young the white line in the front of the eye
is scarcely discernible, and the white bands on the bill are
absent.

Great Auk.

(*See frontispiece.*)

THE Great Auk (*Alca impennis*, Linnæus), which is now
undoubtedly an extinct species, formerly inhabited the North
Atlantic basin. The adult has the general colour of the
plumage similar to that of the razorbill, but its large size,
and the oval patch of white reaching from the eye to the
base of the bill, at once furnishes a means of identification.
Length, about thirty-two inches.

Fulmar Petrel.

JOHN DUNCAN

(*See preceding page.*)

THE Fulmar Petrel (*Fulmarus glacialis*, Linnæus) is a resident in the British Isles, and is also found in the Arctic and Sub-Arctic Regions of the North Atlantic.   The adult male has the back, upper part of wings, and tail, pearl-grey; quills, dusky; remainder of plumage, white; irides, brown; bill, greenish-yellow; legs and feet, light grey.   Length, from nineteen to twenty inches.   The female resembles the male. The young in first plumage are similar to the adults, but are a trifle less in size.   This species is subject to variation in the colour of the plumage.

Capped Petrel.

JOHN DUNCAN

THE Capped Petrel (*Œstrelata hæsitata*, Kuhl) is a very rare irregular migrant to the British Islands, and is supposed to inhabit the South Atlantic. The adult has the crown of the head and nape, black; upper parts, deep brown; forehead, cheeks, neck, rump, and upper tail-coverts, white; central rectrices, mostly blackish, with the basal part more or less white and edged broadly with brown; under plumage, white; bill, black; legs and feet, dull yellowish. Length, about sixteen inches.

Cape Petrel.

JOHN DUNCAN

THE Cape Petrel (*Daption capense*, Linnæus), a very
rare straggler to the British Isles, is found inhabiting
the South Atlantic Ocean and South Pacific Ocean. This
petrel can readily be identified by the sooty-coloured head,
the variegated dusky and white of the upper plumage, and
the pure white of the under parts.

Collared Petrel.

THE Collared Petrel (*Estrelata torquata*, Macgillivray) is a
very rare abnormal migrant to the British Islands.

Bulwer's Petrel.

BULWER'S Petrel (*Bulweria columbina*, Moquin-Tandon) is
a very rare abnormal spring migrant to the British Isles,
and is found inhabiting the Northern Circum-tropical seas.
The adult has the general colouration of the plumage
nearly uniform sooty-black; greater wing-coverts margined
with paler; the tail consists of twelve feathers, and is
wedge-shaped; irides, dark hazel; bill, black; legs and
toes, deep reddish-brown; webs, dark brown.   Length,
about ten inches.

Wilson's Petrel.

JOHN DUNCAN.

WILSON'S Petrel (*Oceanites Wilsoni*, Bonaparte) is a rare straggler to the British Islands, and is found inhabiting the Southern Seas. The adult has the general colour of the plumage sooty-black; head and neck, lighter; wings and tail (the latter nearly square), darker; wing-coverts and inner secondaries, edged with greyish; rump, upper tail-coverts, and patch on thighs, white; base of outer rectrices, marked with white; irides, deep hazel; bill, black; legs and feet, black; webs, yellow at their bases. Length, about seven inches.

Leach's Fork-tailed Petrel.

Leach's Fork-tailed Petrel (*Procellaria Leachi*, Temminck) is a resident in the British Seas, and is also found inhabiting the North Atlantic and North Pacific.   The adult male has the general colour of the plumage sooty-brown ; feathers of upper tail-coverts, white ; some of under tail-coverts, also white ; tail, much forked ; irides, brown ; bill, legs, and feet, dusky-black.   Length, about seven inches.   The female is similar to the male.   The young are said to resemble the adults.

Stormy Petrel.

THE Stormy Petrel (*Procellaria pelagica*, Linnæus) is a resident in the British Seas. It inhabits the Atlantic coasts of Europe, and also the Mediterranean. The adult male has the whole of the plumage black, with the exception of a slight edging of whitish on wing-coverts, upper tail-coverts, and sides of rump, which are white; irides, brown; bill, legs, and feet, black. Length, about five inches. The female is like the male. The young show less whitish on the wings, and not so much white on the sides of the rump.

Madeira Storm Petrel.

THE Madeira Storm Petrel (*Oceanodroma cryptoleucura*) has occurred in the British Isles.

White-bellied Storm Petrel.

THE White-bellied Storm Petrel (*Pelagodroma marina*) is now included in the list of British birds.

White=throated Grey Petrel.

THE White-throated Grey Petrel (*Orstrelata brevipes*) is also included in the list of British birds.

Manx Shearwater.

THE Manx Shearwater (*Puffinus anglorum*, Temminck) is a resident species in the British Seas, and is also found frequenting the whole of the North Atlantic, but is not so numerous on the Western as on the Eastern side. The adult has the upper parts nearly black; sides of head and neck, mottled with ashy-black; under parts, white; irides, brown; bill, brownish; legs and feet, yellowish flesh colour. Length, about fourteen inches. The female resembles the male. The young are similar to the adults, but occasional examples show a more sooty colouration on the under parts.

Sooty Shearwater.

JOHN DUNCAN.

THE Sooty Shearwater (*Puffinus griseus*, Gmelin) is a casual
visitant during our summer to the British Islands, the
breeding area being in the Southern Seas.   The adult has
the general colouration of the upper plumage sooty-brown,
with lighter margins to most of the feathers of back, scapulars, and wing-coverts; primaries and tail, blackish; chin,
whitish; rest of under parts, greyish-brown, several of the
feathers with paler centres; irides, deep brown; bill, dark
horn colour; legs and feet, brown; webs, paler.   Length,
from sixteen to seventeen inches.   It is said that the young
closely resemble the adults.

Dusky Shearwater.

JOHN DUNCAN.

THE Dusky Shearwater (*Puffinus obscurus*, Gmelin) is an exceedingly rare abnormal spring migrant to the British Islands, and inhabits the Tropic Seas. This Shearwater in adult plumage closely resembles *Puffinus anglorum* in colouration, but is less in size, and the upper parts are darker. Length, about eleven inches.

Great Shearwater.

THE Great Shearwater (*Puffinus major*, F. Faber) is an irregular visitor to the British coasts in autumn. It also inhabits the Atlantic Ocean. The adult is, above, ashy-grey, most of the feathers margined with paler; neck, white, nearly all round; wings and tail, blackish; under parts, white, with pale brown markings on middle of belly and thighs; irides, dark brown; bill, dark horn colour; legs and toes, brown; webs, dingy flesh colour. Length, from eighteen to nineteen inches.

Levant's Shearwater.

LEVANT'S Shearwater (*Puffinus yelkouanus*) is an exceedingly rare visitor to the British Isles.

Red=throated Diver.

THE Red-throated Diver (*Colymbus septentrionalis*, Linnæus) is a resident in the British Isles, and also inhabits the Northern Palearctic Region, as well as the Nearctic Region. The adult in nuptial plumage has the forehead, sides of head, and neck, grey; crown, nape, and back of neck, streaked longitudinally with white and black; centre portion of throat, rusty red; back, scapulars, lesser and greater wing-coverts, dusky-brown, marked more or less on the back with white spots; quills, brownish; under parts, glossy white; irides, brown; bill, black; legs, black; centre of webs, tinged with yellowish.   Length, from

twenty-three to twenty-five inches. The female is similar
to the male, but is a little less in size. The young have
the feathers on the upper parts fringed with white.

Black=throated Diver.

JOHN DUNCAN

THE Black-throated Diver (*Colymbus Arcticus*, Linnæus)
is a resident in the British Islands, and is also found in
the Northern Palearctic and Nearctic Regions.   The
adult in summer dress has the upper parts blackish, with
bars and spots of white ; crown and hind-neck, ashy-grey ;
chin and throat, black, glossed with purple, and crossed
by a band of black and white streaks ; sides of neck and
chest, longitudinally banded with white and black ; below

chest, white ; irides, red ; bill, black ; legs and feet, blackish-brown. Length, about twenty-six inches. The female is similar to the male, but smaller. The young have the upper parts dull brownish-black, with light margins on some of the feathers ; and the scapulars and wing-coverts are spotted sparingly with white ; under parts, white.

Great Northern Diver.

JOHN DUNCAN.

THE Great Northern Diver (*Colymbus glacialis*, Linnæus) is a fairly common autumn and winter migrant to the British Isles, and is found inhabiting North-Western Europe and North-Eastern America.   The adult in summer has the

head and neck jet black, glossed with purple and green;
throat, with two white bands streaked longitudinally with
black lines; sides of chest, white, streaked with black;
upper plumage, mostly black, variegated with rows of white
patches and spots; under parts, white; sides and flanks,
blackish, marked with white; irides, crimson; bill, black;
legs and feet, olive-black.   Length, from thirty-two to
thirty-three inches.   The male is larger than the female.
In winter the throat-bands are absent.

White=billed Diver.

JOHN DUNCAN.

THE White-billed Diver (*Colymbus Adamsi*, Gray) is a rare
nomadic winter migrant to the British Isles.   It inhabits
the Circumpolar Region.   The adult resembles the great
northern diver, but in the former the bill is yellowish-white
at all seasons and is deeper; besides, the under mandible is

much up-curved from the angle, and the white streaks on the upper throat-band number only six, whilst there are ten on the under.   In *Colymbus glacialis* the white streaks on the upper throat-band number about twelve, and on the lower eighteen.   The white-billed diver is also larger in size than the latter.

Great Crested Grebe.

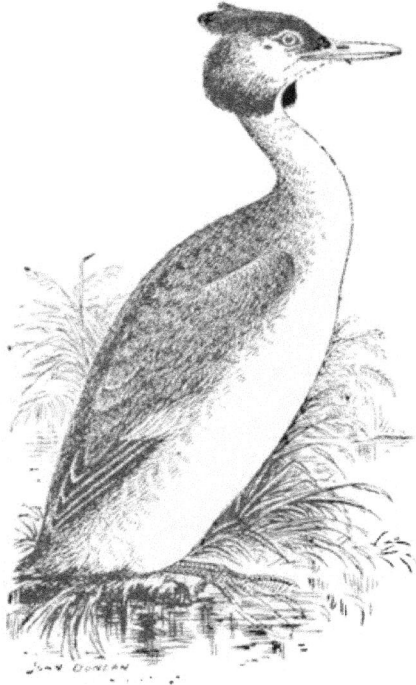

The Great Crested Grebe (*Podiceps cristatus*, Linnæus) is a somewhat locally distributed resident species in the British Isles, and is also found residing in most portions of the

Central and Southern Palearctic Region, moving southward to Southern Africa, India, Australia, and New Zealand. The adult male in summer has the general colour of the upper parts brown, tinged with greyish, with paler margins; wings, crossed with a white band; forehead, crown, and crests, greyish-brown; eye-streak, cheeks, and chin, white; frill round the neck, chestnut, margined on hinder part with black; under parts, silky white; flanks, brownish, with mottlings of rufous; irides, red; bill, red; legs and feet, olive. Length, from twenty to twenty-two inches.

RED-NECKED GREBE.

THE Red-necked Grebe (*Podiceps rubricollis*, Gmelin) is a regular visitor to the British Isles during winter. It also occurs in most parts of Eastern and Northern Europe. The adult in summer dress has the crown, nape, and hind part of neck, blackish; cheeks, chin, and throat, ashy-grey, margined with white; upper parts, blackish-brown; secondaries, white; front part of neck, rich chestnut; abdomen, white; irides, straw colour; bill, blackish, yellow at base; legs and feet, deep olive. Length, from seventeen to eighteen inches. The adult female in summer is similar in colour, but slightly less in size. The adult in winter has the throat white, tinged with grey. The young are similar. (P. 412.)

Red=necked Grebe.

Sclavonian Grebe.

John Duncan

THE Sclavonian Grebe (*Podiceps cornutus*, Gmelin) is a straggling autumn and winter migrant to the British Islands, and inhabits the Northern Nearctic and Palearctic Regions. The adult male in summer has the forehead, crown, and neck-ruff, glossy black; lores, brownish-red; occipital tufts, and streak behind the eye, yellowish-chestnut; upper plumage, deep brown; feathers, margined with paler; quills, dusky; secondaries, mostly white or dusky; fore part, sides of neck, and flanks, yellowish-chestnut ; remainder of under parts, glossy white ; irides, crimson ; bill, straight, and black ; under base and tip, red ; bare part between base of bill and eye, crimson ; legs and feet, olive ; webs, paler.   Length, from twelve to thirteen inches.   The adults in winter lack the neck-ruff and ear-tufts.

Black=necked Grebe.

THE Black-necked Grebe (*Podiceps nigricollis*, Brehm) is only a stray visitor to the British Islands in spring, and

occasionally in autumn.   It is also found in the Southern
Palearctic Region, and throughout the whole of Africa.
The adult in summer has the upper parts dark brown ;
secondaries, nearly all white ; four innermost primaries also
show much white ; head and neck, black, tinged with olive
on the crown ; behind the eye is a triangular patch of
elongated feathers of a rich golden chestnut colour ; under
parts, white, washed on the flanks with dull chestnut ;
irides, red ; bill, black, and slightly recurved ; legs and feet,
olive-green.   Length, about twelve inches.   The female is
similar to the male.

LITTLE GREBE.

THE Little Grebe (*Podiceps minor*, Brisson) is a resident in
the British Islands, in most parts of temperate Europe, and
Asia.   It is also met with in the Malay Archipelago,
Northern Australia, Madagascar, and Africa.   The adult in
breeding plumage has the head, neck, and upper parts,
dark brown ; secondaries, marked with white ; sides of face
and fore part of neck, chestnut ; abdomen, greyish-white ;
sides, mottled with deep brown ; irides, brown ; bill, black,
lighter at tip ; legs and feet, deep olive.   Length, from
eight to ten inches.   The female is similar to the male.
The young show more brown above.   (P. 416.)

Little Grebe.

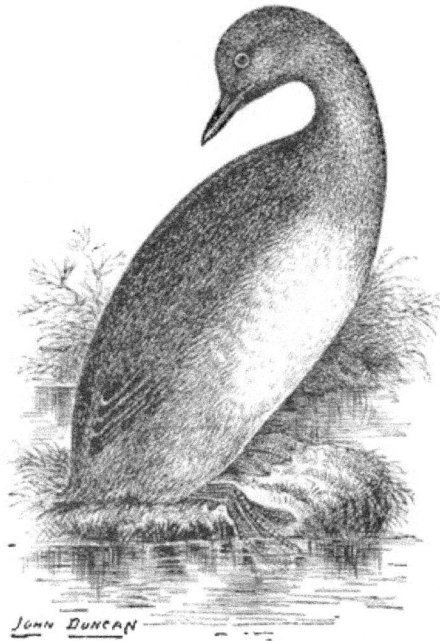

JOHN DUNCAN

Corn Crake.

THE Corn Crake (*Crex pratensis*, Bechstein) is a well-known
visitor to our shores in summer, and occasionally examples
are met with in winter. It is also found throughout the
Western Palearctic Region. The adult male is, above,
dark brown, feathers edged with pale ferruginous; wing-
coverts, chestnut; primaries, reddish-brown; throat, white;
patch over eyes, cheeks, and breast, light ash; centre of
abdomen, nearly white; flanks and under tail-coverts,
marked with broad bars of yellowish-rufous and whitish
mottlings; irides, light brown; bill, legs, and feet, pale

27

pinkish brown.   Length, about eleven inches.   The female
is a trifle less in size than the male, and the colouration is
duller.

Spotteð Crake.

THE Spotted Crake (*Crex porzana*, Linnæus) visits the
British Isles in the summer, and a few individuals spend
the winter with us.   It is also found inhabiting temperate
Europe and Western Asia.   In winter it retires southwards
to Africa and India.   The adult male has the crown of the
head dark brown; eye-stripe, cheeks, and throat, greyish;
breast, brownish, with white spots; abdomen, greyish;
flanks, marked with bars of brown and white; upper parts,
mostly olive-brown, streaked with darker brown, and spotted
minutely with white; irides, hazel; bill, yellow; base,
orange; legs and feet, olive.   Length, from eight to nine
inches.   The female is duller in colour, and a little less in
size than the male.   The young are said to have the throat
white and the spots fainter.

Little Crake.

THE Little Crake (*Crex parva*, Scopoli) is a rare irregular spring and autumn migrant to the British Islands, and inhabits the Western Palearctic Region. The adult male has the crown of the head, hind neck, and upper parts, olivaceous-brown; centre of back, streaked broadly with black, and spotted here and there with white; both webs of quills, clove-brown; forehead, cheeks, sides of neck, and remainder of under parts, bluish-grey, with small spots of white on the thighs, vent, and under tail-coverts; rectrices, with dark centres; irides, red; bill, green, reddish at base; legs and feet, green. Length, about eight inches. In the adult female the eye is surrounded with light grey; throat, whitish; neck, breast, and abdomen, washed with buff; flanks, brown, with transverse bars of white.

Baillon's Crake.

BAILLON'S Crake (*Porzana Bailloni*, Vieill) is an exceedingly rare visitor to the British Isles, and is stated to have nested twice with us. This bird is also found in the Southern Palearctic Region and throughout Africa. The adult male has the crown, neck, and upper parts chiefly rufous-brown, marked with white and black streaks and spots; sides of head, throat, and breast, bluish-slate; outer web of first primary, white; flanks and under tail-coverts, with boldly defined black and white bars; irides, red; bill, yellowish-green; legs and feet, flesh colour, tinged with brown. Length, about seven inches. The adult female is lighter above than the male; under parts, browner.

Water=rail.

THE Water-rail (*Rallus aquaticus*, Linnæus) is a generally distributed resident species throughout the British Isles, and is also found in most parts of the Western Palearctic Region.   The adult male has the top of the head and upper parts chiefly reddish-brown, streaked in centre of feathers with blackish; chin, nearly white; cheeks, neck, breast, and abdomen, lead colour; flanks, barred transversely with black and white; primaries, brownish; tail, dark brown; vent, pale buffish-white; irides, hazel; bill, red; legs and feet, flesh colour, tinged with brown.   Length, from ten to eleven inches.   The adult female closely resembles the male, but is a little less in size.   The young in first plumage have the under parts buffish-white, with bars of dark brown on the flanks.

Carolina Rail.

Snow Sc   JOHN DUNCAN

THE Carolina Rail (*Crex Carolina*, Linnæus) has one
British record. It is a migratory species, and in summer
is found inhabiting the Northern United States, as far
north as lat. 62°; in winter it occurs in the Southern
States, Mexico, Central America, and the West Indies.

Waterhen.

THE Waterhen (*Gallinula chloropus*, Linnæus) is a widely distributed resident species in the British Islands, and occurs in most parts of Europe, Asia, and Africa. The adult has the head, nape, and upper parts, dark slate-grey; under parts, dusky; flanks, brownish, with a few white streaks; under tail-coverts, white; horny shield on forehead, scarlet; irides, hazel; bill, scarlet, yellow at tip; legs, tinged with yellow and green; tibia, scarlet. Length, about thirteen inches. The female is said to be richer in colour than the male. The young have the horny shield of a greenish tinge.

Common Coot.

THE Common Coot (*Fulica atra*, Linnæus) is a resident
in the British Islands, moving about locally after the
nesting season.   It is also found distributed over most
portions of Europe and temperate Asia.   The adult has
the upper parts dark slate-grey; secondaries, tipped with
white; under parts, slate-grey; horny shield on forehead,
white; irides, crimson; bill, pinkish at base, white at tip;
legs and feet, dark green.   Length, from fifteen to sixteen
inches.   The female is similar to the male.   The young
have the horny shield less developed, and show white on
the throat.

Ring=dove.

THE Ring-dove (*Columba palumbus*, Linnæus) is a resident species in the British Islands, and, with the exception of the extreme North, inhabits the whole of the Western Palearctic Region, also ranging through Western Asia. The adult has the upper parts bluish-ash; rump, paler; lower portion of neck and breast, bluish-grey, suffused with purplish-red and green; sides of neck have a conspicuous patch of white; wings, brownish; wing-coverts, edged with white; tail, blackish; centre of belly, light grey; irides, light yellow; bill, red at base, towards tip yellowish; legs and feet, crimson. Length, from fifteen to seventeen inches. The adult female has the neck spot smaller, and the general plumage is duller. The young show no neck spot.

Stock=dove.

THE Stock-dove (*Columba œnas*, Linnæus) is a resident in
the British Islands, and also occurs throughout the Western
Palearctic Region, and ranges into Asia.   The adult male
is readily distinguished from the ring-dove by its smaller
size and absence at all times of the white neck patch, and
from the wings showing no white.   The wing bar rudimen-
tary, rump, and general plumage, greyish-blue; fore part
of breast, tinged with vinous purple; sides of neck have
purple and green reflections; irides, red; bill, whitish, red
at base; legs and feet, crimson.   Length, about thirteen
inches.   The adult female is duller in plumage and less in
size.   The young have the plumage browner, and lack the
sheen on the neck.

Rock-dove.

THE Rock-dove (*Columba livia*, Brisson) is another resident in the British Isles, and is also found in most parts of the Western Palearctic Region, with the exception of the Far North. The adult male can at once be distinguished from the stock-dove by each wing having two broad black transverse bars, and lower back and rump being white. Irides, orange; bill, blackish; legs and feet, red. Length, from eleven to thirteen inches. The female is similar to the male, but is less in size. The young show no sheen on the plumage.

Turtle-dove.

THE Turtle-dove (*Turtur auritus*, Gray), which is one of our summer migrants, is met with in the Western Palearctic Region. The adult male has the head, neck, breast, and back, light wood-brown, tinged with pearl-grey: on either side of neck is a conspicuous patch of black feathers, with white tips; scapulars and wing-coverts, dark-brown, margined with cinnamon-brown; primaries, brownish-black; two centre feathers of tail, clove-brown; remainder of feathers have white tips; breast, pale vinous-red; belly, white; flanks, lavender; under tail-coverts, white; irides, rufous; bill, brown; legs and feet, crimson. Length, eleven or twelve inches. The adult female is similar to the male, but the plumage is duller in colour. The young in first plumage want the patch on the neck.

Eastern Turtle-dove.

THE Eastern Turtle-dove (*Turtur Orientalis*, Latham) is a very rare irregular autumn migrant to the British Isles; it is also found inhabiting the North-western Oriental Region and South-eastern Palearctic Region.   This dove may readily be recognised by the "broad chestnut-brown margins of its dark central scapulars and tertials"

Passenger Pigeon.

THE Passenger Pigeon (*Ectopistes migratorius*, Linnæus) is a migratory species, which has occurred several times in the British Isles, and in summer is found inhabiting North

America. The adult male has the head and upper parts bluish-grey, with spots of black on the wings; primaries, brownish-black, edged with grey; breast, cinnamon-rufous, changing into pale vinous on the remainder of under parts; under tail-coverts, white.

Pallas's Sand Grouse.

PALLAS's Sand-grouse (*Syrrhaptes paradoxus*, Pallas) is an exceptionally rare visitor at intervals to the British Islands. The bird is also found inhabiting the Tartar Steppes of Asia, occurring more or less throughout the greater part of Europe. The adult male has the upper parts warm ochreous-buff, with bars of black; primaries, long, pointed, and greyish; head, yellowish-buff; two central tail-feathers, long and pointed; neck and breast, buff, tinged with grey, and crossed by a mottled band; a broad black band runs across the abdomen; vent, white; irides, dark brown; feet, feathered to the toes. Length, from thirteen to fifteen inches. The adult female is said to have no black band across the breast, and is less in size than the male.

Common Quail.

THE Common Quail (*Coturnix communis*, Bonnaterre) is a widely distributed species in summer in this country, and examples are often met with during the autumn and winter. It is also found inhabiting the Palearctic Region and the whole of Africa, and migrates from Northern Europe during winter. The adult male has the general colouration of the plumage brown, varied on the upper parts with streaks of buff; eye-streak, creamy-white; throat, rufous, having a double crescent of black below; lower part of neck and breast, light rufous; belly, pale buff; flanks, buffish-red, striped longitudinally with buff; irides, brown; bill, legs, and feet, brown, with a yellowish tinge. Length, about seven inches.

Red=legged Partridge.

THE Red-legged Partridge (*Caccabis rufa*, Linnæus) was introduced into this country upwards of one hundred years ago. It inhabits the western portion of Europe. The adult male can easily be distinguished from the common partridge by the black gorget and the flanks, which are barred with black and chestnut ; irides, hazel ; legs, red. Length, from twelve to thirteen inches. The adult female is less in size than the male, and the plumage is duller. The young in first plumage are browner below than the adults, and the gorget is not so fully developed.

Common Partridge.

THE Common Partridge (*Perdix cinerea*, Brisson) is a generally distributed resident in the British Isles, and also inhabits the temperate portions of Europe and Western Asia. A detailed description of this well-known species is hardly needed, but the fact may be mentioned that the mature male can be distinguished from the female by the rich dark brown marks on the lower breast, those on the female being much less developed. The young have no dark patches on the belly.

28

Pheasant.

THE Pheasant (*Phasianus colchicus*, Linnæus) is an introduced species, and was originally an inhabitant of South-
Eastern Europe and Asia Minor. A description of this
common and well-known bird is hardly required.

Capercailzie.

JOHN DUNCAN.

THE Capercailzie (*Tetrao urogallus*, Linnæus), which became extinct about a hundred years ago, has since been
introduced, and is now fairly abundant in the central
portion of Scotland. It also inhabits the Northern Palearctic Region. The general colouration of the plumage of
the adult male is dark slate-grey; scapulars, wing-coverts,
and primaries have fine vermiculations of whitish; under

parts, marked with white; breast, shaded with green; tail, rounded and black, with spots of white; irides, brown; bare spots above eyes, bright red; bill, whitish.   Length, from thirty-three to thirty-six inches.   The female, which is less than the male, is of a yellowish chestnut, with variegations of black, brown, rufous, and white.

Black Grouse.

THE Black Grouse (*Tetrao tetrix*, Linnæus) is a resident in this country; it also inhabits Northern Europe, and is met with in some parts of Central as well as Southern

Europe. The adult male is so well known that a brief description is quite sufficient. The general colour is bluish-black, tinged with brown on upper parts; white bar across the wings; tail, black, outermost rectrices curved outwards; irides, hazel; wattles above eyes, scarlet. Length, from twenty to twenty-three inches. The adult female is less than the male, and is similar in colouration to the female capercailzie; tail, not forked. The young resemble the female.

Red Grouse.

THE Red Grouse (*Lagopus scoticus*, Brisson) is confined to the British Isles. The adult male in spring is principally chestnut-brown, the greater portion of the feathers being vermiculated, barred, and speckled with black; abdomen, greyish-white; irides, hazel; feathers on legs and feet, greyish-white; wattles over eyes, rich scarlet. This species varies much in colouration. Length, from fifteen to sixteen inches. The adult female is less in size,

and the general colour is much paler. The young resemble
the adults, but show more white on the head and belly.

Ptarmigan.

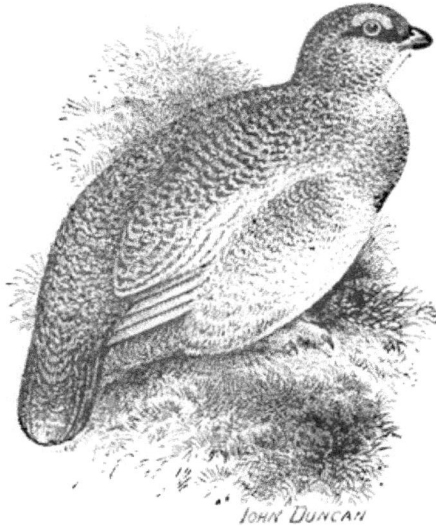

JOHN DUNCAN

THE Ptarmigan (*Lagopus mutus*, Montin) is a resident in the
British Islands, and is also found inhabiting the mountainous portions of Europe and Central Asia. The adult
male in summer plumage has the chin and under parts
below the chest, white; wings, white; tail, black; central
feathers have narrow tips of white; remainder of plumage,
almost black, with fine vermiculations of buffish-brown;
lores, black; head and neck, marked here and there with
white; irides, hazel; wattles above eyes, scarlet; bill and
claws, dark brown. Length, about fifteen or sixteen inches.
The female's general colour is rusty-yellow, barred and
otherwise marked with blackish.

APPENDIX.

A list of species which are said to have occurred in the British Isles.

Alpine Chough (*Pyrrhocorax Alpinus*).
Red-winged Starling (*Agelæus phœniceus*).
Rusty Grakle (*Scolecophagus ferrugineus*).
Meadow Starling (*Sturnella magna*).
Gold-vented Bulbul (*Pycnonotus capensis*).
South African Serin (*Serinus canicollis*).
Yellow-rumped Seed-eater (*Serinus icterus*).
Nonpariel Finch (*Cyanospiza ciris*).
White-throated Sparrow (*Zonotrichia albicollis*).
Ruby-crowned Wren (*Regulus calendula*).
American Robin (*Turdus migratorius*).
White-collared Flycatcher (*Musicapa collaris*).
Red-rumped Swallow (*Hirundo rufula*).
American Tree Swallow (*Tachycineta bicolor*).
Blue-tailed Bee-eater (*Merops Philippinus*).
Abyssinian Roller (*Coracias leucocephalus*).
Indian Roller (*Coracias Indicus*).
Jugger Falcon (*Falco jugger*).
American Kestrel (*Falco sparverius*).
Black-winged Kite (*Elanus cæruleus*).
Desert Buzzard (*Buteo desertorum*).
Red-tailed Buzzard (*Buteo borealis*).
Red-shouldered Buzzard (*Buteo lineatus*).
Trumpeter Swan (*Cygnus buccinator*).

American Swan (*Cygnus Americanus*).
Bar-headed Goose (*Anser Indicus*).
Chinese Goose (*Anser cygnoidus*).
Canada Goose (*Bernicia Canadensis*).
Egyptian Goose (*Chenalopex Ægyptiaca*).
Spur-winged Goose (*Plectropterus gambensis*).
Ring-necked Duck (*Fuligula collaris*).
Little Green Heron (*Butorides virescens*).
Soudan Crane (*Grus pavonina*).
Andalucian Hemipode (*Turnix sylvatica*).
Marsh Sandpiper (*Totanus stagnatalis*).
Bridled Tern (*Sterna anæstheta*).
Pied-billed Grebe (*Podilymbus podiceps*).
Albatross (*Diomedea melanophrys*).
Purple Gallinule (*Porphyrio cæruleus*).
Green-backed Gallinule (*Porphyrio smaragdonotus*).
Martinique Gallinule (*Porphyrio Martinicus*).

INDEX.

ACCENTOR, ALPINE, 126
,, Hedge, 125
Alpine Pipit, 59
American Bittern, 283
,, Golden Plover, 312
,, Goshawk, 206
,, Pectoral Sandpiper, 335
,, Stint, 338
,, Teal, 239
,, White-winged Cross-bill, 14
,, Wigeon, 235
Aquatic Warbler, 99
Arctic Blue-throated Robin, 115
,, Tern, 376
Asiatic Golden Plover, 312
Auk, Great, 392
,, Little, 389
Avocet, Common, 315

BAILLON'S CRAKE, 420
Barn Owl, 156
,, Swallow, 134
Barred Warbler, 93
Bar-tailed Godwit, 347
Bartram's Sandpiper, 324
Bean Goose, 224
Bearded Titmouse, 69
Bee-eater, 151
Belted Kingfisher, 155
Bernacle Goose, 231
Bewick's Swan, 221
Bittern, 280
,, American, 283
,, Little, 281
Blackbird, 108
Black Grouse, 436
,, Guillemot, 388
,, Kite, 195
· ,, Redstart, 118

Black Stork, 279
,, Tern, 383
,, ,, White-winged, 384
Blackcap Warbler, 91
Black-backed Gull, Great, 359
,, ,, Lesser, 361
Black-bellied Dipper, 128
Black-billed Cuckoo, 144
Black-headed Gull, 363
,, ,, Great, 364
,, ,, Mediterra-nean, 365
Black-necked Grebe, 414
Black-tailed Godwit, 346
Black-throated Diver, 407
,, Ouzel, 110
,, Wheatear, 121
Blue Titmouse, 74
Blue-headed Wagtail, 55
Blue-throated Robin, Arctic, 115
Blue-winged Teal, 239
Blyth's Willow Wren, 83
Bonaparte's Gull, 366
,, Sandpiper, 332
Brambling, 24
Brent Goose, 229
,, ,, White-bellied, 230
Broad-billed Sandpiper, 334
Brunnich's Guillemot, 387
Buff-backed Heron, 286
Buff-breasted Sandpiper, 336
Buffel-headed Duck, 260
Buffon's Skua, 354
Bullfinch, 20
,, Greater, 20
Bulwer's Petrel, 397
Bunting, Black-headed, 42
,, Brandt's Siberian, 45
,, Cirl, 39
,, Corn, 38

Bunting, Lapland, 36
,, Little, 43
,, Ortolan, 41
,, Reed, 37
,, Rustic, 44
,, Snow, 35
,, Yellow, 40
Bustard, Great, 295
,, Little, 296
,, Macqueen's, 298
Buzzard, Common, 201
,, Honey, 199
,, Rough-legged, 203

CALANDRA LARK, 50
Canary, 29
Cape Petrel, 396
Capercailzie, 435
Capped Petrel, 395
Carolina Rail, 422
Caspian Sand Plover, 309
,, Tern, 378
Chaffinch, 23
Chiffchaff, 81
Chough, Common, 6
Cirl Bunting, 39
Coal Titmouse, 73
Collared Petrel, 396
Common Avocet, 315
,, Buzzard, 201
,, Eider, 261
,, Guillemot, 385
,, Gull, 362
,, Heron, 283
,, Jay, 7
,, Kingfisher, 154
,, Kite, 193
,, Nightjar, 149
,, Nuthatch, 68
,, Partridge, 433
,, Pratincole, 302
,, Quail, 431
,, Sandpiper, 329
,, Scoter, 256
,, Snipe, 350
,, Stilt, 312
,, Swift, 146
,, Teal, 238
,, Tern, 375
,, Wren, 129

Coot, Common, 424
Cormorant, 216
Corn Bunting, 38
Coues' Redpole, 34
Courser, Cream-coloured, 300
Crake, Baillon's, 420
,, Corn, 417
,, Little, 419
,, Spotted, 418
Crane, Common, 293
,, Demoiselle, 293
Creeper, Common, 64
,, Wall, 65
Crested Lark, 48
,, Titmouse, 71
Crossbill, American White-
winged, 14
,, Common, 13
,, European White-
winged, 15
,, Parrot, 16
Crow, Carrion, 2
,, Hooded, 3
Cuckoo, 142
,, Black-billed, 144
,, Great Spotted, 143
,, Yellow-billed, 145
Curlew, Common, 315
,, Sandpiper, 325
,, Stone, 298

DARTFORD WARBLER, 87
emoiselle Crane, 293
Desert Wheatear, 122
Dipper, 127
,, Black-bellied, 128
Diver, Black-throated, 407
,, Great Northern, 408
,, Red-throated, 406
,, White-billed, 409
Dotterel, 301
Dove, Ring, 425
,, Rock, 427
,, Stock, 426
,, Turtle, 428
Duck, Buffel-headed, 260
,, Harlequin, 254
,, Long-tailed, 255
,, Pintail, 241
,, Tufted, 250

Dunlin, 341
Dusky Redshank, 344
,,   Shearwater, 404

EAGLE, GOLDEN, 187
,,   Owl, 170
,,   Spotted, 189
,,   White-tailed, 191
Eastern Turtle-dove, 429
Egret, Great White, 291
,,   Little, 289
Egyptian Nightjar, 150
,,   Vulture, 173
Eider, Common, 261
,,   King, 264
,,   Steller's, 262
Eskimo Whimbrel, 318
European White-winged Cross-
bill, 15

FALCON, ICELAND JER, 177
,,   Peregrine, 178
,,   Scandinavian Jer, 177
,,   White Jer, 174
Fieldfare, 107
Finch, Scarlet Rose, 18
,,   Serin, 28
Firecrest, 66
Flamingo, 273
Flycatcher, Pied, 132
,,   Red-breasted, 133
,,   Spotted, 131
Fork-tailed Petrel, Leach's, 399
Fulmar Petrel, 394

GADWALL, 240
Gannet, 218
Garden Warbler, 90
Garganey, 242
Glossy Ibis, 273
Godwit, Bar-tailed, 347
,,   Black-tailed, 346
Goldcrest, 67
Golden Eagle, 187
,,   Eye, 252
,,   Oriole, 12
,,   Plover, 310
,,   ,,   American, 312
,,   ,,   Asiatic, 312
Goldfinch, 26

Goosander, 266
Goose, Bean, 224
,,   Bernacle, 231
,,   Brent, 229
,,   Grey-Lag, 223
,,   Lesser Snow, 222
,,   ,,   White-fronted, 227
,,   Pink-footed, 225
,,   Red-breasted, 232
,,   White-bellied Brent, 230
,,   White-fronted, 226
Goshawk, 204
,,   American, 206
Grasshopper Warbler, 101
Great Auk, 392
,,   Bustard, 295
,,   Northern Diver, 408
,,   Reed Warbler, 98
,,   Shearwater, 405
,,   Skua, 353
,,   Snipe, 349
,,   Spotted Cuckoo, 143
,,   ,,   Woodpecker, 141
,,   Titmouse, 75
,,   White Egret, 291
Greater Bullfinch, 20
,,   Ringed Plover, 306
Grebe, Black-necked, 414
,,   Great Crested, 410
,,   Little, 415
,,   Red-necked, 411
,,   Sclavonian, 413
Green Sandpiper, 328
,,   Woodpecker, 139
Greenfinch, 25
Greenland Redpole, 34
Greenshank, 345
Grey Petrel, White-throated, 401
,,   Phalarope, 320
,,   Plover, 311
,,   Shrike, Great, 80
,,   ,,   Lesser, 79
,,   ,,   Pallas's, 81
,,   Wagtail, 54
Griffon Vulture, 172
Grosbeak, Pine, 17
Ground Thrush, Siberian, 112
,,   ,,   White's, 111
Grouse, Black, 436
,,   Red, 437

Guillemot, Black, 388
  ,,   Brunnich's, 387
  ,,   Common, 385
  ,,   Ringed, 386
Gull, Black-headed, 363
  ,,   Bonaparte's, 366
  ,,   Common, 362
  ,,   Glaucous, 370
  ,,   Great Black-backed, 359
  ,,   Black-headed, 364
  ,,   Herring, 359
  ,,   Iceland, 371
  ,,   Ivory, 372
  ,,   Lesser Black-backed, 361
  ,,   Little, 367
  ,,   Mediterranean Black-
         headed, 365
  ,,   Ross's, 369
  ,,   Sabine's, 368
Gull-billed Tern, 379

Harlequin Duck, 254
Harrier, Hen, 211
  ,,   Marsh, 213
  ,,   Montagu's, 209
Hawfinch, 19
Hawk Owl, 165
  ,,   Sparrow, 205
Hedge Accentor, 125
Heron, Buff-backed, 286
  ,,   Common, 283
  ,,   Night, 285
  ,,   Purple, 289
  ,,   Squacco, 287
Herring Gull, 359
Hobby, 179
  ,,   Orange-legged, 180
Holboell's Redpole, 34
Honey Buzzard, 199
Hooded Crow, 3
  ,,   Merganser, 265
Hooper Swan, 220
Hoopoe, 153
House Martin, 135
  ,,   Sparrow, 21

Ibis, Glossy, 273
Iceland Gull, 371
  ,,   Jer-Falcon, 177
Icterine Warbler, 103

Isabelline Wheatear, 120
Ivory Gull, 372

Jackdaw, 5
Jack Snipe, 351
Jay, Common, 7
Jer-Falcon, Iceland, 177
  ,,   Scandinavian, 177
  ,,   White, 174

Kentish Sand Plover, 308
Kestrel, 183
  ,,   Lesser, 184
Killdeer Plover, 305
King Eider, 264
Kingfisher, Belted, 155
  ,,   Common, 154
Kite, Black, 195
  ,,   Common, 193
  ,,   Swallow-tailed, 197
Kittiwake, 357
Knot, 342

Lapland Bunting, 36
Lapwing, 303
  ,,   Sociable, 304
Lark, Calandra, 50
  ,,   Crested, 48
  ,,   Shore, 51
  ,,   Short-toed, 47
  ,,   Sky, 45
  ,,   White-winged, 49
  ,,   Wood, 46
Leach's Fork-tailed-Petrel, 399
Lesser Black-backed Gull, 361
  ,,   Grey Shrike, 79
  ,,   Kestrel, 184
  ,,   Redpole, 32
  ,,   Snow Goose, 222
  ,,   Spotted Woodpecker, 140
  ,,   Tern, 377
  ,,   White-fronted Goose, 227
  ,,   Whitethroat, 88
Levant's Shearwater, 405
Linnet, 30
Little Auk, 389
  ,,   Bittern, 281
  ,,   Bunting, 43
  ,,   Bustard, 296
  ,,   Crake, 419
  ,,   Egret, 289

Little Grebe, 415
,, Gull, 367
,, Owl, 163
,, Ringed Plover, 307
,, Stint, 337
Long-eared Owl, 160
Long-tailed Duck, 255
,, Titmouse, 70

MACQUEEN'S BUSTARD, 298
Madeira Storm Petrel, 401
Magpie, 8
Mallard, 245
Manx Shearwater, 402
Marsh Harrier, 213
,, Titmouse, 72
,, Warbler, 94
Martin, House, 135
,, Purple, 137
,, Sand, 136
Meadow Pipit, 58
Mealy Redpole, 33
Mediterranean Black-headed Gull, 365
Melodious Warbler, 100
Merganser, Hooded, 265
,, Red-breasted, 269
Merlin, 182
Missel Thrush, 105
Montagu's Harrier, 209
Mute Swan, 219

NEEDLE-TAILED SWIFT, 148
Night Heron, 285
Nightingale, 116
Nightjar, Common, 149
,, Egyptian, 150
,, Red-necked, 150
Noddy Tern, 381
Nutcracker, 9
Nuthatch, Common, 68

ORANGE-LEGGED HOBBY, 180
Oriole, Golden, 12
Orphean Warbler, 92
Ortolan Bunting, 41
Osprey, 215
Ouzel, Black-throated, 110
,, Ring, 109
Owl, Barn, 156

Owl, Eagle, 170
,, Hawk, 165
,, Little, 163
,, Long-eared, 160
,, Scops, 168
,, Short-eared, 158
,, Snowy, 164
,, Tengmalm's, 161
,, Wood, 157
Oyster Catcher, 321

PALLAS'S GREY SHRIKE, 81
,, Sand-grouse, 430
,, Willow Wren, 82
Parrot Crossbill, 16
Partridge, Common, 433
,, Red-legged, 432
Passenger Pigeon, 429
Pectoral Sandpiper, American, 335
,, Siberian, 336
Peregrine Falcon, 178
Petrel, Bulwer's, 397
,, Cape, 396
,, Capped, 395
,, Collared, 396
,, Fulmar, 394
,, Leach's Fork-tailed, 399
,, Madeira Storm, 401
,, Stormy, 400
,, White-bellied Storm, 401
,, White-throated Grey, 401
,, Wilson's, 398
Phalarope, Grey, 320
,, Red-necked, 319
Pheasant, 434
Pied Flycatcher, 132
,, Wagtail, 52
Pine Grosbeak, 17
Pink-footed Goose, 225
Pintail Duck, 241
Pipit, Alpine, 59
,, Meadow, 58
,, Red-throated, 63
,, Richard's, 62
,, Rock, 60
,, Tawny, 61
,, Tree, 57
Plover, American Golden, 312
,, Asiatic Golden, 312
,, Caspian Sand, 309

Plover, Golden, 310
,,   Greater Ringed, 306
,,   Grey, 311
,,   Kentish Sand, 308
,,   Killdeer, 305
,,   Little Ringed, 307
,,   Ringed, 307
Pochard, 246
,,   Red-crested, 247
,,   White-eyed, 248
Pomatorhine Skua, 356
Pratincole, Common, 302
Ptarmigan, 438
Puffin, 390
Purple Heron, 289
,,   Martin, 137
,,   Sandpiper, 333

QUAIL, COMMON, 431

RAIL, CAROLINA, 422
,,   Water, 421
Raven, 1
Razorbill, 391
Red Grouse, 437
Red-backed Shrike, 77
Red-breasted Flycatcher, 133
,,   Goose, 232
,,   Merganser, 269
,,   Snipe, 352
Red-crested Pochard, 247
Red-legged Partridge, 432
Red-necked Grebe, 411
,,   Nightjar, 150
,,   Phalarope, 319
Red-throated Diver, 406
,,   Pipit, 63
Redpole, Coues', 34
,,   Greenland, 34
,,   Holboell's, 34
,,   Lesser, 32
,,   Mealy, 33
Redshank, 343
,,   Dusky, 344
Redstart, 117
,,   Black, 118
Redwing, 106
Reed Bunting, 37
,,   Warbler, 96
,,   ,,   Great, 98

Richard's Pipit, 62
Richardson's Skua, 355
Ring-dove, 425
Ring Ouzel, 109
Ringed Guillemot, 386
,,   Plover, 307
,,   ,,   Greater, 306
,,   ,,   Little, 307
Robin, 114
,,   Arctic Blue-throated, 115
Rock Pipit, 60
Rock-dove, 427
Rock-Thrush, 113
Roller, 152
Rook, 4
Roseate Tern, 374
Rose-coloured Starling, 11
Rose Finch, Scarlet, 18
Ross's Gull, 369
Rough-legged Buzzard, 203
Ruddy Sheldrake, 234
Ruff, 322
Rufous Warbler, 95
Rustic Bunting, 44

SABINE'S GULL, 368
St. Kilda Wren, 130
Sanderling, 340
Sand Martin, 136
,,   Plover, Caspian, 309
,,   ,,   Kentish, 308
Sand-grouse, Pallas's, 430
Sandpiper, American Pectoral, 335
,,   Bartram's, 324
,,   Bonaparte's, 332
,,   Broad-billed, 334
,,   Buff-breasted, 336
,,   Common, 329
,,   Curlew, 325
,,   Green, 328
,,   Purple, 333
,,   Siberian Pectoral, 336
,,   Solitary, 331
,,   Spotted, 326
,,   Wood, 326
,,   Yellow-legged, 330
Sandwich Tern, 373
Savi's Warbler, 102
Scandinavian Jer-Falcon, 177

Scarlet Rose Finch, 18
Scaup, 251
Sclavonian Grebe, 413
Scops Owl, 168
Scoter, Common, 256
    ,,   Surf, 259
    ,,   Velvet, 258
Sedge Warbler, 97
Serin Finch, 28
Shag, 217
Shearwater, Dusky, 404
    ,,   Great, 405
    ,,   Levant's, 405
    ,,   Manx, 402
    ,,   Sooty, 403
Sheldrake, 233
    ,,   Ruddy, 234
Shore-Lark, 51
Short-eared Owl, 158
Short-toed Lark, 47
Shoveller, 244
Shrike, Great Grey, 80
    ,,   Lesser Grey, 79
    ,,   Pallas's Grey, 81
    ,,   Red-backed, 77
    ,,   Woodchat, 78
Siberian Bunting, Brandt's, 45
    ,,   Ground Thrush, 112
Siskin, 27
Skua, Buffon's, 354
    ,,   Great, 353
    ,,   Pomatorhine, 356
    ,,   Richardson's, 355
Sky-Lark, 45
Smew, 271
Snipe, Common, 350
    ,,   Great, 349
    ,,   Jack, 351
    ,,   Red-breasted, 352
Snow Bunting, 35
    ,,   Goose, Lesser, 222
Snowy Owl, 164
Sociable Lapwing, 304
Song Thrush, 104
Sooty Tern, 380
Sparrow Hawk, 205
    ,,   House, 21
    ,,   Tree, 22
Spoonbill, 276
Spotted Crake, 418

Spotted Cuckoo, Great, 143
    ,,   Eagle, 189
    ,,   Flycatcher, 131
    ,,   Woodpecker, Great, 141
    ,,   ,,   Lesser, 140
Squacco Heron, 287
Starling, 10
    ,,   Rose-coloured, 11
Steller's Eider, 262
Stilt, Common, 312
Stint, American, 338
    ,,   Little, 337
    ,,   Temminck's, 339
Stock-dove, 426
Stonechat, 124
Stone Curlew, 298
Stork, Black, 279
    ,,   White, 276
Storm Petrel, Madeira, 401
    ,,   ,,   White-bellied, 401
Stormy Petrel, 400
Subalpine Warbler, 93
Swallow, Barn, 134
Swallow-tailed Kite, 197
Swan, Bewick's, 221
    ,,   Hooper, 220
    ,,   Mute, 219
Swift, Common, 146
    ,,   Needle-tailed, 148
    ,,   White-bellied, 147

TAWNY PIPIT, 61
Teal, American, 239
    ,,   Blue-winged, 239
    ,,   Common, 238
Temminck's Stint, 339
Tengmalm's Owl, 161
Tern, Arctic, 376
    ,,   Black, 383
    ,,   Caspian, 378
    ,,   Common, 375
    ,,   Gull-billed, 379
    ,,   Lesser, 377
    ,,   Noddy, 381
    ,,   Roseate, 374
    ,,   Sandwich, 373
    ,,   Sooty, 380
    ,,   Whiskered, 382
    ,,   White-winged Black, 384
Thrush, Missel, 105

Thrush, Rock, 113
,, Siberian Ground, 112
,, Song, 104
,, White's Ground, 111
Titmouse, Bearded, 69
,, Blue, 74
,, Coal, 73
,, Crested, 71
,, Great, 75
,, Long-tailed, 70
,, Marsh, 72
Tree Pipit, 57
,, Sparrow, 22
Tufted Duck, 250
Turnstone, 323
Turtle-dove, 428
,, Eastern, 429
Twite, 31

Velvet Scoter, 258
Vulture, Egyptian, 173
,, Griffon, 172

Wagtail, Blue-headed, 55
,, Grey, 54
,, Pied, 52
,, White, 53
,, Yellow, 56
Wall-Creeper, 65
Warbler, Aquatic, 99
,, Barred, 93
,, Blackcap, 91
,, Dartford, 87
,, Garden, 90
,, Grasshopper, 101
,, Great Reed, 98
,, Icterine, 103
,, Marsh, 94
,, Melodious, 100
,, Orphean, 92
,, Reed, 96
,, Rufous, 95
,, Savi's, 102
,, Sedge, 97
,, Subalpine, 93
Waterhen, 423
Water-rail, 421
Waxwing, 76
Wheatear, 119
,, Black-throated, 121
,, Desert, 122

Wheatear, Isabelline, 120
Whimbrel, 317
,, Eskimo, 318
Whinchat, 123
Whiskered Tern, 382
White Egret, Great, 291
,, Jer-Falcon, 174
,, Stork, 276
White's Ground Thrush, 111
Whitethroat, 89
,, Lesser, 88
White-bellied Brent Goose, 230
,, Storm Petrel, 401
,, Swift, 147
White-billed Diver, 409
White-eyed Pochard, 248
White-fronted Goose, 226
,, ,, Lesser, 227
White-tailed Eagle, 191
White-throated Grey Petrel, 401
White-winged Black Tern, 384
,, Crossbill, American, 14
,, Crossbill, European, 15
,, Lark, 49
Wigeon, 236
,, American, 235
Wilson's Petrel, 398
Wood Owl, 157
,, Sandpiper, 326
Wood-Lark, 46
Woodchat Shrike, 78
Woodcock, 348
Woodpecker, Great Spotted, 141
,, Green, 139
,, Lesser Spotted, 140
Wren, Blyth's Willow, 83
,, Common, 129
,, Pallas's Willow, 82
,, St. Kilda, 130
,, Willow, 85
,, Wood, 86
,, Yellow-browed Willow, 84
Wryneck, 138

Yellow Bunting, 40
,, Wagtail, 56
Yellow-billed Cuckoo, 145
Yellow-browed Willow Wren, 84
Yellow-legged Sandpiper, 330

LIST OF SUBSCRIBERS.

A

Adams, Edward, Edenville, Stocksfield

Adamson, Horatia A., 29 Percy Gardens, Tynemouth

Adamson, Miss Charlotte S., North Jesmond

Addison, T., Green Street, South Shields

Affleck, Thomas, 16 West Black Street, East Jarrow

Affleck, W., Brown's Buildings, nr. Chester-le-Street

Alexander, T. D., 27 St. George's Terrace, Newcastle

Allan, T. & G., Blackett Street, Newcastle

Allan, William, Grainger Street, Newcastle

Allan, W. Tone, 6 Thorney Terrace, South Shields

Allhusen, Alfred, Beadnell, Chathill

Anderson, George, Little Harle Tower, Newcastle

Anderson, M., Dinnington Colliery

Anne, Major, Army and Navy Club, London

Appleby, John, South West Farm, Monkseaton

Armstrong, W. A. Watson, Cragside, Rothbury

Arnison, W. C., M.D., 4 Fenham Terrace, Newcastle

Avery, John, Christon Bank, R.S.O.

Avery, W. A., Cragside, Rothbury

B

Bailes, Thomas, Jesmond Gardens, Newcastle

Balmbra, John, Lisburn Street, Alnwick

Barlow, Joseph, 134 Northumberland Street, Newcastle

Barrow, F., Rothbury, Northumberland

Bell, Henry H., 174 Rye Hill, Newcastle

Bell, T. J., Old Hall, Cleadon, near Sunderland

Bell, William Storey, 2 Quality Row. Cambois

Bewick, Luke, 110 James Street, Blaydon

Blair, Dr. Ernest, 4 Thorney Terrace, South Shields

Blair, Robert, Harton, South Shields

Bland, J. D. W., 61 South Street, Durham Road, Gateshead

Blanshard, Paul, Sandyford Park, Jesmond

29

Blunt, H. P., 1 Elysium Place, Bensham, Gateshead

Boazman, John, 25 Quayside, Newcastle

Boiston, James, High Burn House, Heworth

Bolam, Edwin, 231 Monday Street, Newcastle

Bolam, George, 55 Castle Gate, Berwick

Braithwaite, John, Bank of England, Newcastle

Brewis, James, 21 Elmwood Street, Sunderland

Brown, Joseph B., 61 Hotspur Street, Newcastle

Brown, Dr. R., Blaydon-on-Tyne

Brown, Robert, Little Houghton, Lesbury

Brown, R. H., Melton Constable Inn, Seaton Sluice, Northumberland

Browne & Browne, 103 Grey Street, Newcastle

Brumell, Dr. Arthur, Pethgate House, Morpeth

Brumell, George, The Willows, Morpeth

Butler, Alice, Chipchase, Wark-on-Tyne

C

Cackett, Jas. T., 113 Osborne Road, Newcastle

Cameron, John, 2 Hall Terrace, Gateshead

Carr, Arthur D., The Manor House, Alnmouth

Carr, Rev. Charles B., Vicarage, Long Framlington, Morpeth

Carr, Edward, 30 Howick Street, Alnwick

Carr, Rev. O. C., All Saints' Vicarage, Newcastle

Carr-Ellison, H. G., 25 Eldon Place, Newcastle

Carr-Ellison, John R., Hedgeley, Alnwick

Charlton, Henry, J.P., 1 Millfield Terrace, Gateshead

Charlton, Joseph, 3 Victoria Place, Monkseaton

Charlton, Thomas, Haddrick's Mill, Newcastle

Cicester, Ernest R., The Palace, Chichester

Clark, G. D. Atkinson, Belford Hall, Belford

Clark, Isaac, Jun., 133 Thomas Terrace, Blaydon

Clark, James, 82 Rye Hill, Newcastle

Cole, R. P., 2 Dene Crescent, Walker

Cooke, H., Benwell Grove, Newcastle

Connell, Frederick, Edward Road, Whitley

Cowen, Mrs. John A., Blaydon Burn House, Blaydon

Cowen, Joseph, Stella Hall, Blaydon

Crawhall, Geo. E., 38 Eldon Street, Newcastle

Creighton, Robert, Wansbeck Street, Morpeth

Cresswell, John, Rothbury House, Heaton, Newcastle

Crowe, George, 25 Rowley Street, Blyth

Cunningham, J., Ship Inn, Felling Shore

D

Davidson, Robert, Assistant Overseer, Monkseaton

Davie, John K., 119 Burt Terrace, Gateshead

Dawson, John, Monkseaton

Dawson, Matt. H., 41 Elswick Row, Newcastle

Dawson, Thomas, Milbourne Arms, Holywell Village, Seaton Delaval

Denison, Joseph, Solicitor, 45 Sanderson Road, Newcastle

Dick, Thomas, Manchester Street, Morpeth

Dixon, Sir Raylton, Gunnergate Hall, Marton, R.S O., Yorks

Dobinson, Thomas, Monkseaton

Donkin, Edward, North Field, Thorne, near Doncaster

Dowson, John, Thorp Avenue, Morpeth

Druery, G., Clarendon Terrace, South Shields

Dryden, Robert, Monkseaton

Dudgeon, John Scott, Long Newton, St. Boswell's, N.B.

Duncan, R., 2 Landsdowne Terrace, Gosforth

Duncan, Thomas Elliott, 2 Henry Street, Coatham, Redcar

Duncan, W., 40 Green Street, South Shields

Dunn, Nathaniel, Shilbottle Colliery, Lesbury, R.S.O.

Dunn, Wm., Dolphin Inn, Spital Dene, Tynemouth

Durey, John, Forest Hall

E

Edmonds, Temple, 64 Linskill Terrace, North Shields

Elliott, E., 70 Crown Street, Newcastle

Emley, Frederick, Ravenshill, Durham Road, Gateshead

Emmerson, John, 6 Rectory Terrace, Gosforth

Ewen, George, 11 Duke Street, Whitley

F

Farthing, A. P., 8 Leazes Crescent, Newcastle

Fawcett, William, Blyth

Fenwick, F., Eshott Hall, Felton, Northumberland

Fenwick, J. C. J., M.D., Long Framlington, Morpeth

Ferguson, William, 15 Prudhoe Street, Newcastle

Forster, Charles, Farnley Hill, Corbridge

Forster, C. Frank, Southill, Chester-le-Street

Forster, Fred. E., 32 Grainger Street West, Newcastle

Forster, G. B., Farnley Hill, Corbridge

Forster, Henry Joseph, Bradley Cottage, Wolsingham, Co. Durham

Forster, John, Tyne Printing Works, Newcastle

Forster, W. C., 13 Grainger Street West, Newcastle

Fox, Rev. H. E., The Croft, Lytton Grove, Putney, London

Franklin, W. E., Mosley Street, Newcastle

Fry, Charles Rutter, Parkside, Darlington

G

Gibson, J. P., Photographer, Hexham

Gibson, Thomas, 9 Amberley Street, Sunderland

Glaholm, F. J., 98 Grey Street, Newcastle

Grace, H. W., Hallgarth Hall, Winlaton

Grace, William Percy, Whickham

Graham, Samuel, 107 High Park Road, Newcastle

Green, R. Y., 11 Lovaine Crescent, Newcastle

Greene, W. T., M.A., M.D., F.Z.S., Iveagh Lodge, Belvedere, Kent

Greenwell, G. C., F.G.S., M. Inst. C.E., Duffield, Derby

Grey, Sir Edward, Falloden, Chathill, Northumberland

Grubb, J. H., 5 Thorney Terrace, South Shields

H

Halliday, T., Myrtle Cottage, Low Fell, Gateshead

Hassell, C., Tynemouth

Haswell, George H., Handsworth, Birmingham

Haswell, Lieut.-Col. F. R. A., Monkseaton

Haworth, Rev. J., St. Hilda's College, Durham

Hedley, John T., Rainham, Beckenham, Kent

Henderson, G. E., 16 Framlington Place, Newcastle

Henderson, John W., 8 Sydenham Terrace, South Shields

Henderson, W. J., Whitburn Gardens, Sunderland

Herdman, Thomas, Westgate Chambers, Cross Street, Newcastle, and 6 Gordon Square, Whitley

Heslop, Jas., Broomhaugh, Riding Mill

Heslop, R. Oliver, 12 Akenside Hill, Newcastle

Hicks, William Barnes, Holly Cottage, Monkseaton

Higginbottom, A. H., Simmonley, Jesmond, Newcastle

Hills, Thos. E., Red Lion, Earsdon

Hills, William, Black Horse Hotel, Monkseaton

Hodges, Charles C., Hexham

Hodgson, J. Duncan, Linton Villa, Grainger Park Road, Newcastle

Holmes, Ald. Richard Henry, J.P., 54 Rye Hill, Newcastle

Holmes, Wm. Henry, Wellburn, Newcastle

Howse, Richard, Museum, Barras Bridge, Newcastle

Hoyle, Theodore, Burdon Buildings, Grainger Street West, Newcastle

Hudson, Thomas, Vine Cottage, North Shields

Hughes, Geo. P., F.R.G.S., etc., Middleton Hall, Wooler

Hunter, Henry, Old Hartley, Northumberland

Hunter, W., 3 Catherine Terrace, Whitley

Hutchinson, John, 19 Alexandra Road, Gateshead

Hutchinson, Jos., The College, Durham

Hutchinson, Thomas, 1 Clarence Crescent, Whitley

Hutchinson, Wm. W., 12 Marine Avenue, Whitley

I

Irving, John A., West Fell, Corbridge

Irwin, Charles, Osborne House, Tynemouth

J

Jackson, John, 26 High Bridge, Newcastle

Jackson, S. F., Post Office, 38 Coatsworth Road, Gateshead

James, J. J., 8 Newgate Street, Morpeth

Jaques, George, Monkseaton

Jobling, James, Bank, Morpeth

Jobling, Robert Lee, 48 Sixth Avenue, Heaton, Newcastle

Johnson, J., Cuthbert Street, South Shields

Johnson, Thomas, Tyne House, Golding's Hill, Loughton, Essex

Johnston, D. R., 91 South·View West, Heaton, Newcastle

Johnston, Robert, 39 Green Street, South Shields

Johnston, W., 3 High Bridge, Newcastle

Joicey, Sir James, Longhirst, Morpeth

K

Keen, John, 16 Choppington Street, Newcastle

Kightley, Alfred, 43 Queen's Terrace, Jesmond, Newcastle

Kinghorn, Mrs. Jane, Druridge, Widdrington, Acklington

Kinloch, Thomas, Station Master, Woodburn, Northumberland

Kirsopp, John, Bank, Blyth

L

Lamb, Edmund, Borden Wood, Liphook, Hants

Lamb, Friend, Ashmore House, Sunderland

Lambert, Thomas, Jun., 38 Beverley Terrace, Cullercoats

Langley, P. S., 5 Kingston Square, Hull

Latimer, William, Berney Hill Lodge

Lee, Mrs. S., 14 Countess Street, Whitley

Lincoln, E. H., Market Place, South Shields

Linton, Mrs. M., Crook

Lowery, D. A., 285 Westgate Road, Newcastle

Lowrey, Jos., F.R.G.S., The Hermitage, Loughton, Essex

Luckley, R., 9 Nesham Street, Newcastle

Lupton, Banister, Beechcroft, Gosforth

M

Macdonald, E., 51 Falconar Street, Newcastle

Mackay, Matthew, 8 Milton Street, Newcastle

McLean, Hugh, Main Street, Corbridge

Makepeace, J. B., Lilac House, Springwell

Marshall, J., Dairy House, Seaton Delaval

Martin, Alexander, Broomhouse Lodge, Beal

Maudlen, Wm., Alma Place, North Shields

Mawson, Swan & Morgan, Grey Street, Newcastle

May, John, 37 Granville Park, Lewisham, S.E.

Merivale, John H., Togston Hall, Acklington

Middlemiss, Mr., 7 Dundas Street, Monkwearmouth, Sunderland

Middleton, Sir Arthur, Belsay Castle, Newcastle

Middleton, J. T., Banbury Road, Brackley, Northants

Miller, James, 46 Jesmond Road, Newcastle

Miller, John E., Middleham, Yorks

Minshull, A., Monkseaton

Minto, W. J., Bird Hill, Whickham

Mitcalfe, Stanley, Tynemouth

Mitford, Lieut.-Colonel John, V.D., Glanton Pyke, Glanton, R.S.O.

Moore, John, Greylands, Marine Parade, Penarth, Glam.

Morris, Thomas, 6 Newton Street, Bensham, Gateshead

Muir, D. K., L.R.C.P. & S., etc., Byker Terrace, Walker

Munro, Alex., 2 St. James' Terrace, Newcastle

N

Nicholson, H., Monkseaton

Nicholson, John, Byegate Farm, Monkseaton

Northern Counties' Educational, Trading, and School Furnishing Co., Ltd., Athenæum Street, Sunderland

North of England School Furnishing Co., 34 Fawcett Street, Sunderland

O

Oliver, Dr. Thomas, 7 Ellison Place, Newcastle

Oliver, Thomas, Ennerdale, Fernwood Road, Jesmond, Newcastle

P

Pape, Victor, 5 Lovaine Row, Newcastle

Paynter, Henry Augustus, Freelands, Alnwick

Pease, John W., Pendower, Newcastle

Pepper, Fred J., Handsworth, Birmingham

Pickering, Thomas, Tyneholme, Osborne Road, Newcastle

Pigg, Joseph, 24 Newgate Street, Newcastle

Plummer, B., Rose Hill, Low Fell

Poole, F. John, St. John Street, Ashbourne, Derbyshire

Porteus & Co., R. J., Grainger Street West, Newcastle

Potts, Matthew, Monkseaton West Farm

Pow, Robert, Whitley

Powell, M., 4 Union Place, Gateshead

Proudlock, Joseph, Milburn Terrace, Seaton Delaval

Purves, John, Tweed House, Archbold Terrace, Newcastle

Purvis, A., South View, South Shields

Pyke, T., Librarian, Public Library, South Shields

R

Ravensworth, Earl of, Ravensworth Castle, Gateshead

Reed, Joseph, 3 Benton Terrace, Newcastle

Reed, R. B., Springfield, Forest Hall, Newcastle

Reid, E. O., 15 North Terrace, Newcastle

Reid, John, Blue House Farm, Fulwell, near Sunderland

Reid, Wm. B., Cross House, Newcastle

Reid, W. B., Jun., Cross House, Newcastle

Richardson, Albert, & Co., Complete School Furnishers, Durham

Richardson, Thos., Bates' Cottages, Holywell

Richardson, T., 37 Gloucester Street, Newcastle

Richardson, Wm., Bates' Cottages, Seaton Delaval

Ridley, The Hon. Justice Edward, 48 Lennox Gardens, London

Ritson, U. A., Jesmond Gardens, Newcastle

Roberts, John S., 79 Warwick Street, Newcastle

Robertson, W. L., 18 Meldon Terrace, South Shields

Robinson, R., 15 Ninth Avenue, Heaton, Newcastle

Rogers, David, 134 Barras Bridge, Newcastle

Rutherford, John V. W., M.R.C.S., Briarwood, Newcastle

S

Saunders, Edwin, Blyth

Scott, Walter, Beauclerc, Riding Mill

Scott, Mason T., Beauclerc, Riding Mill

Scott, Charles T., The Lodge, Ardsley, near Barnsley

Scott, William Martin, Leeds Steel Works, Leeds

Scott, George, Westoe, South Shields

Scott, John, Jun., 66 Redheugh Terrace, Gateshead

Sherwood, Henry, 1 Grosvenor Villas, Jesmond, Newcastle

Smart, W. P., 38 Countess Avenue, Whitley

Smith, Roger B., Sea View House, Monkseaton

Smith, Thomas, Avenue Villa, Durham

Smith, W. J., Flass, Durham

Smith, William, Gunnerton, Barrasford

Snowdon, A., Hutton Rudby, Yarm, Yorkshire

Stephens, Rev. Thomas, Horsley Vicarage, Otterburn

Stephenson, J. A., Inglesyde, Low Fell

Stewart, Thomas, Allotment, Shiremoor

Stokoe, H., Creekside House, Beckton, North Woolwich

Stothard, Jas. Tone, 1 St. Cuthbert's Terrace, North Shields

Straker, Jos. H., Howden Dene, Corbridge

Strang, Wm. Wailes, 53 Beverley Terrace, Cullercoats

Stuart, John, Innkeeper, Newton-by-the-Sea, Chathill

Swallow, Thomas, 37 Bell Terrace, Newcastle

Swanston, Wm., 15 Victoria Square, Newcastle

T

Taylor, Rev. Edward James, F.S.A., St. Cuthbert's, Durham

Taylor, W. B., Northumberland Park, North Shields

Thompson, George H., Alnwick

Thompson, Joseph L., Westholme Hall, Winston-on-Tees, via Darlington

Towers, Edward, 4 Latimer Street, Tynemouth

Turnbull, Jos. Wm., Belmont, Riding Mill

Tweddle, Thomas, Jun., 12 Church Street, Winlaton

Tyson, Wilson, 43 Lovaine Place, Newcastle

W

Waddington, Thos., 3 Park Lane, Leeds

Wade, J. R., Laygate Lane, South Shields

Wakefield, Edwin, 203 Coatsworth Road, Gateshead

Walker, John D., Solicitor, 21 Pilgrim Street, Newcastle

Ward, M., Hill Street, South Shields

Ward, W., 60 Cuthbert Street, South Shields

Warden, Thos. M., 21 Windermere Street, Gateshead

Watson, Frank, 51 Hartington Street, Newcastle

Watson, Robert Spence, Bensham Grove, Gateshead

Webb, William, 23 Newgate Street, Morpeth

Webster, Anna, 20 Chester Street, Newcastle

Wedderburn, J. Ventress, Whitley

Welford, R., Thornfield, Gosforth

Whitehorn, Augustus, Egremont House, Whitley

Whitfield, Robert, 5 Bloomfield Terrace, Gateshead

Whittaker, Frederick W., 57 Prince's Dock, Liverpool

Williamson, David A., 28 Percy Gardens, Tynemouth

Williamson, Thos., Lovaine House, North Shields

Willits, John, Cullercoats, Whitley

Wilkinson, C. J., 13 Ewell Road, Surbiton, Surrey

Wilkinson, William C., 12 Dacre Street, Morpeth

Wilson, D. F., 2 Maritime Place, Morpeth

Wilson, Edward, Monkhouse, Tynemouth

Wilson, E. F., Lough House, Morpeth

Wilson, Fenwick, Marden, Whitley

Woodward, Frederick R., 13 Woodbine Street, Gateshead

Wraith, T. W., 5 Richmond Terrace, Gateshead

Wright, Andrew, Deaf and Dumb Institution, Newcastle

Y

Yellowley, A., 80 Brodrick Street, South Shields

Yeoman, A. R., M.A., Ingleside, Sanderson Road, Newcastle

Young, George, Oswald House, Morpeth

Young, J. P., 158 Rye Hill, Newcastle

Webster, Anna, 20 Chester Street, Newcastle

Wedderburn, J. Ventress, Whitley

Welford, R., Thornfield, Gosforth

Whitehorn, Augustus, Egremont House, Whitley

Whitfield, Robert, 5 Bloomfield Terrace, Gateshead

Whittaker, Frederick W., 57 Prince's Dock, Liverpool

Williamson, David A., 28 Percy Gardens, Tynemouth

Williamson, Thos., Lovaine House, North Shields

Willits, John, Cullercoats, Whitley

Wilkinson, C. J., 13 Ewell Road, Surbiton, Surrey

Wilkinson, William C., 12 Dacre Street, Morpeth

Wilson, D. F., 2 Maritime Place, Morpeth

Wilson, Edward, Monkhouse, Tynemouth

Wilson, E. F., Lough House, Morpeth

Wilson, Fenwick, Marden, Whitley

Woodward, Frederick R., 13 Woodbine Street, Gateshead

Wraith, T. W., 5 Richmond Terrace, Gateshead

Wright, Andrew, Deaf and Dumb Institution, Newcastle

Y

Yellowley, A., 80 Brodrick Street, South Shields

Yeoman, A. R., M.A., Ingleside, Sanderson Road, Newcastle

Young, George, Oswald House, Morpeth

Young, J. P., 158 Rye Hill, Newcastle

Subscribers omitted from the preceding List.

Appleby, T. D., 29 Woodbine Road, Gosforth

Blenkinsop, W. R., 8 Eslington Terrace, Jesmond, Newcastle

Brumell, F., Town Clerk, Morpeth

Foy, Charles, Market Place, Wokingham

Hodgson, David, The Hollins, Newby (West), Carlisle

Jenkins, H. T., Westbury, Wiltshire

Nicholson, G., 65 Hartington Street, Newcastle

www.ingramcontent.com/pod-product-compliance
Lightning Source LLC
Chambersburg PA
CBHW020902210326
41598CB00018B/1747